High Performance and Hybrid Manufacturing Processes

High Performance and Hybrid Manufacturing Processes

Guest Editor

Ali Abdelhafeez Hassan

MDPI

Basel • Beijing • Wuhan • Barcelona • Belgrade • Novi Sad • Cluj • Manchester

Guest Editor
Ali Abdelhafeez Hassan
Institute of Engineering,
Computing and Advanced
Manufacturing
University of Cumbria
Carlisle
UK

Editorial Office
MDPI AG
Grosspeteranlage 5
4052 Basel, Switzerland

This is a reprint of the Special Issue, published open access by the journal *Machines* (ISSN 2075-1702), freely accessible at: https://www.mdpi.com/journal/machines/special_issues/hybridMachining_Processes.

For citation purposes, cite each article independently as indicated on the article page online and as indicated below:

Lastname, A.A.; Lastname, B.B. Article Title. *Journal Name* **Year**, *Volume Number*, Page Range.

ISBN 978-3-7258-6362-4 (Hbk)
ISBN 978-3-7258-6363-1 (PDF)
https://doi.org/10.3390/books978-3-7258-6363-1

Contents

About the Editor

Ali Abdelhafeez Hassan

Ali Abdelhafeez Hassan is an innovative and research-driven mechanical engineer with over 15 years of academic and industrial experience across the UK and Egypt. He has specialized in smart manufacturing and digital twinning, with a strong track record of interdisciplinary collaboration, curriculum development, and mentoring.

He is recognized for impactful contributions to aerospace, robotics, biomedical, and renewable energy sectors through cutting-edge research, patents, and publications in collaboration with major UK manufacturers like Rolls-Royce, Airbus Operations Ltd., GKN Aerospace, and many more.

Preface

The manufacturing landscape is undergoing a profound transformation driven by the demand for higher efficiency, precision, and sustainability. 'High Performance and Hybrid Manufacturing Processes' brings together advanced concepts and emerging technologies that are shaping the future of production systems. This reprint aims to provide readers with a consolidated resource that captures the latest developments, research insights, and practical applications in this dynamic field.

Hybrid manufacturing—integrating additive and subtractive techniques—has opened new possibilities for complex geometries, material optimization, and cost-effective production. Coupled with high performance processes such as laser-based machining, advanced forming, and intelligent automation, these innovations are redefining traditional paradigms and enabling manufacturers to meet the challenges of Industry 4.0.

This collection serves as a valuable reference for researchers, engineers, and practitioners seeking to understand and implement cutting-edge manufacturing strategies. By bridging theory and practice, it highlights not only the technological advancements but also the interdisciplinary nature of modern manufacturing, where materials science, digital technologies, and process engineering converge.

We hope this reprint inspires further exploration and collaboration in developing sustainable, efficient, and high-quality manufacturing solutions for the future.

Ali Abdelhafeez Hassan
Guest Editor

machines

MDPI

Article

DLP of Translucent Alumina: In-Depth Investigation on Slurry Development and Debinding Regimes

Michele De Lisi [1,*], Chang Shu [1], Usama M. Attia [2] and Khamis Essa [1,*]

1 Department of Mechanical Engineering, University of Birmingham, Birmingham B15 2TT, UK
2 The Manufacturing Technology Centre Limited, Ansty Park, Coventry CV7 9JU, UK
* Correspondence: mxd962@student.bham.ac.uk (M.D.L.); k.e.a.essa@bham.ac.uk (K.E.)

Abstract: Exploring the feasibility of producing near-net-shape components with advanced properties and geometrical features via 3D printing has incrementally become the research focus of various studies. Digital light processing (DLP) technology can manufacture complex-structured components for various technical applications. The aims of this research were to investigate Al_2O_3 ceramic slurry preparation procedures to identify the ideal components to add to an in-house-developed ceramic slurry, to determine the optimal DLP printing parameters and conditions while understanding their effect on the green part properties and to evaluate the appropriate debinding regime to achieve fully dense crack-free fired parts capable of exhibiting translucent behaviours. The slurry obtained from the ball-milled powder at 800 rpm for 1 h, together with 2 wt.% BYK-145 as a dispersant and the highest achievable solid loading of 85 wt.%, showed the desired rheological and photopolymerisation properties. Full-factorial design of experiments (DOE) was employed to study the impact of the printing parameters on the density and the dimensions of the samples. Different debinding regimes were investigated and it was proven that the lowest debinding heat rate (0.2 °C/min) and longer holding times helped to reduce defects and promote densification (>99.0%), providing optimal grounds to obtain translucent fired parts.

Keywords: additive manufacturing; DLP; ceramic slurries; debinding; DoE

1. Introduction

Additive manufacturing (AM) represents a possible alternative to conventional production. It is a technique for producing three-dimensional objects by adding individual layers based on a sliced computer-a design (CAD) model. In terms of processed materials, plastics and metals are now the industry leaders, although ceramic additive manufacturing has lately garnered a great deal of attention [1–3]. Indirect and direct ceramic AM are the two most common classifications [4]: the former comprises, among other characteristics, selective laser sintering (SLS), selective laser melting (SLM), and direct inkjet printing (DIP), while the latter includes laminated object manufacturing (LOM) and stereolithography (SLA). SLA is one of the most popular kinds of ceramic-additive-manufacturing techniques and digital light processing (DLP) is one of the most promising processes that originate from the SLA approach while improving it at the same time. The benefit of DLP technology is the more rapid and efficient fabrication of the part, as the projected light source solidifies one layer of slurry at a time [5] rather than a point or a single line.

A crucial objective of AM is the achievement of dimensional precision and reproducibility in the produced parts [6]. For AM of metals, it is common to find extensive research on the effect of printing settings on the part dimensions. For instance, in previous literature, it is possible to find examples of studies where laser powder bed fusion [7–9] and selective laser melting [10,11] were adopted to create complex metallic structures, and exhaustive research on the effect of the process parameters was conducted. However, on the other hand, the influence of AM processing parameters on the final properties of ceramic components has begun to be investigated only more recently and, even though the conducted studies have been providing progressively more information, further investigations can strengthen the existing knowledge base. Regarding slurry-based methods, in particular, the effects of slurry preparation and printing settings of ceramic stereolithography-based methods on the dimensional accuracy and obtained density have not been thoroughly researched when linked to the final properties of the fired samples. Fu et al. [12] investigated the effect of laser power on the dimensional accuracy of the SLA of bulk items and concluded that the length and breadth increased with increasing laser power due to an increase in light scattering. In a separate investigation, the printing parameters were experimentally altered during the production of a gear shape using LCM and it was discovered that the horizontal resolution rose as the exposure energy decreased [13].

The slurries used for ceramic additive manufacturing via DLP generally comprise ceramic filler, binder, photoinitiator, and other additives. The interactions between all of them must be taken into account when considering possible combinations. For instance, while dimensional error in ceramic-based stereolithography is usually due to the dispersion level of the added ceramic particles to UV resin [14,15], a high index of refraction mismatch between ceramic particles and photopolymerisable resins can reduce the cure depth and alter the resolution [15,16], affecting the achievable accuracy. In the existing literature, there are various examples of the selection process of starting slurry components and different methods for preparing ceramic suspensions are presented. The ceramic powder may be utilised as-is [17–21] or the surface of the ceramic particles may be pre-treated with a dispersant prior to the addition to the photopolymerisable resin [22–24]. There is a dearth of information about the photopolymerisation behaviour and rheology, as well as a lucid description of the optimal way of preparation and achievement of a solid loading greater than 40 vol%, which is a key aspect in the development of ceramic slurries able to obtain successfully dense parts [17,25,26]. High solid loadings are also desired, as they enhance mechanical characteristics while decreasing sintering shrinkages. However, the addition of hydrophilic ceramic particles to the hydrophobic resin dramatically increases viscosity, which directly affects the printing feasibility. To obtain a high solid loading and an excellent dispersion of ceramic particles in the resin, it is necessary to add a dispersant to create steric barriers between ceramic particles that compensate for the van der Waals attractive forces and avoid Brownian motion-induced particle collisions [17].

In addition to the preprocessing and printing parameters, the debinding and sintering processes need to be optimised and strict controls must be established to obtain samples free of cracks [27,28], as low solid loadings can result in 70% shrinkage during the thermal postprocessing. In fact, the 3D-printed ceramic green body is subjected to heat treatment methods in order to produce sintered bodies with the required final qualities. During the debinding procedure, the ceramic slurry's binder must be pyrolysed away. As the binder diffuses through the ceramic green body, microcracks are caused by high heating rates and unsuitable holding temperatures and holding durations [29]. When microcracks are not eliminated, their presence during sintering degrades the mechanical characteristics. Therefore, the heating rate and the dwelling temperatures within the temperature range where the binder decomposes must be accurately calculated prior to thermally stabilising the ceramic green body for an appropriate holding period [27]. In addition, a sufficient sintering temperature is necessary, as the ceramic densification process is incompletely attained when an improper heating temperature is used in the sintering process, resulting in a degradation of the material's physical properties. Therefore, optimal sintering temperatures should be investigated appropriately in advance to fully exploit the potentialities of ceramic materials [30].

Recently, several intriguing experiments have been devoted to the fabrication via AM of ceramic components for special applications in micro-electro-mechanical-system [31] or catalyst-substrate manufacturing [32–34]. However, due to the great optical transparency and characteristic ceramic features of transparent ceramics, they are equally suitable for a wide variety of high-performance applications, such as lasers, armours, and lighting.

There is a pressing need to simultaneously lower the weight of transparent armour systems and boost their ballistic protective qualities. When both mechanical strength and abrasion resistance are essential performance characteristics, material options are frequently restricted to a few materials [35–37], such as aluminium oxynitride (AlON) [38–40], sapphire (Al_2O_3) [41], and magnesium aluminate spinel ($MgAl_2O_4$) [42,43]. Due to their great strength and hardness, these three materials have been investigated as possible replacements for ballistic glasses in transparent armour applications for quite some time, and ballistic tests have demonstrated that spinel, AlON, and sapphire exhibit comparable performance. Payne and colleagues [44,45] claimed that transparent yttrium aluminium garnet (YAG) ceramics may be created via a direct ink writing (DIW) approach, but that the subsequent CIP and HIP processes are still required. Wu et al. [46] continued to use the extrusion-based AM method employing copolymer-assisted slurry so that complex-shaped transparent YAG ceramics could be immediately produced. Hostaša et al. [47] manufactured YAG laser ceramics by lithography-based ceramic manufacturing (LCM) to demonstrate the benefit of lithography-based techniques in terms of spatial resolution and product form. Two-photon printing was used to create small transparent YAG objects with nanometric precision in the nanometre and micrometre ranges (TPP) [48]. However, studies on the DLP of translucent alumina starting from an in-house systematically developed slurry and its consecutive post-processing have yet to be explored.

In conclusion, complex interactions between process settings, printer resolution, and composition of the photopolymerizable slurry determine the precision of DLPed components. Although numerous studies have focused on the effects of printing parameters on the performance of manufactured parts, a systematic examination of the effects of slurry preparation and printing parameters on the geometrical properties of DLPed advanced alumina ceramics is necessary. The impact of different ball-milling pre-treatments, diverse slurry components and, hence, curing behaviours and debinding regimes has to be evaluated by analysing the evolution of the density and the geometrical accuracy at various layer thicknesses and energy dosages. This research aims to analyse the ceramic slurry preparation steps and establish optimal printing and debinding settings for achieving translucency in alumina-based structures even in air-sintering conditions. In this study, the influence of the pre-treatment of ceramic particles relative to the as-received powder was investigated, as well as the additions in the ceramic slurry of different components in various proportions. A full-factorial design of experiment (DoE) was used to explore the influence of layer thickness, exposure power, and exposure time on dimensional accuracy. On the basis of a comparison of the printing's dimensional correctness and on the estimation of the density, the optimal printing parameters were determined. Different debinding regimes were then analysed and, finally, samples were sintered. In the end, the entire process was repeated following the now-established optimal route and adding MgO as a sintering aid in a minimal quantity (1000 ppm) without altering the slurry printability, in order to obtain even a slightly translucent sample.

2. Materials and Methods

2.1. *Ceramic Powder Pretreatment*

Alumina was used as the ceramic powder (CT 3000 LS SG, Almatis GmbH, Ludwigshafen, Germany) for the photopolymerisable slurry. The as-received aluminium oxide powder (D50 = 0.92 μm) was ball milled together with 0.8 mm of aluminium oxide grinding media (Industrie Bitossi S.p.A., Spicchio-Sovigliana, Italy) in a 1:1 ball-to-powder ratio (BPR) at different rotating speeds and holding times in a planetary ball milling (HMK-1901, Dandong HMKTest Instrument Co., Ltd., Dandong, Liaoning, China) in order to see if, within a limited window of investigation, any changes in the pretreated powders could be appreciated. Table 1 illustrates the various values of speed and time tested.

Table 1. Ball-milling parameter combinations investigated.

Run	Milling Speed [rpm]	Milling Time [h]
1	400	1
2	400	2
3	400	3
4	600	1
5	600	2
6	600	3
7	800	1
8	800	2
9	800	3

The set of parameters producing the best results was adopted and finally sieved via a 250 μm mesh screen. The particle size distribution of the ceramic powder was measured by laser diffraction particle size analysis (Sympatec GmbH, Clausthal-Zellerfeld, Germany), which was performed in a liquid dispersion consisting of water and two drops of tetrasodium pyrophosphate as a dispersant. Additionally, SEM images were obtained to analyse the shape of the powder (JEOL JSM-6060LV, Tokyo, Japan).

2.2. Slurry Preparation

The resin was prepared by mixing polyfunctional and monofunctional monomers in different proportions together with a dispersant and a plasticizer (20 wt.% to the resin content) to prevent delamination. Diphenyl (2,4,6-trimethylbenzoyl) phosphine oxide (Sigma Aldrich, Gillingham, UK) was added to the resin mixture as a photoinitiator (1.0 wt.% of the photoreactive resin components). A difunctional and a monofunctional polymer were chosen in a proportion of 6:5 and a ceramic loading of 85 wt.% was selected to be implemented. The processes for preparing ceramic slurry were derived from the literature [49].

The first step consisted of blending the resin components using an overhead mixer at room temperature for 30 minutes, followed by the addition of the dispersant and another 30 minutes of stirring. Then, the ceramic powder was gradually added until a solid loading of 85 wt.% was reached. The photoinitiator was subsequently added to the combination, which was then mixed for further 30 min with the mechanical mixer and, finally, the slurry was placed in a turbula mixer for at least 12 h along with 6 mm spherical alumina grinding media (Industrie Bitossi S.p.A., Spicchio-Sovigliana, Italy) with a BPR = 1. Before the printing, the produced slurry was processed for 10 min in a vacuum degasser to eliminate air bubbles and minimise polymerisation inhibition.

Different dispersants, various percentages of dispersants, and diverse solid loadings were explored. Table 2 shows the components and parameters investigated.

Table 2. Investigated slurry preparation parameters.

Parameter				
Dispersant	BYK-111		BYK-145	BYK-9076
Dispersant content	1 wt.%	2 wt.%	3 wt.%	5 wt.%
Solid loading	75 wt.%	80 wt.%	85 wt.%	90 wt.%

First, the dispersant influence was investigated by trying three different dispersants at the same weight–content percentage to the ceramic powder content and the optimal dispersant was then selected. Second, the effect of the chosen dispersant weight content was studied by varying the weight content between 1% and 5% and the dispersant showing the ideal rheology behaviour was selected. Finally, the solid content workability range was explored by considering four different ceramic solid loadings and the optimal one was chosen. To comprehend the rheological properties of the prepared suspension, the rheology of the ceramic suspensions was examined with AR 500 (TA Instruments, New Castle, DE,

USA) using a 40 mm parallel plate geometry in the shear rate window ranging from 0.1 to 300 s^{-1} at a constant temperature of 20.0 °C, in accordance with DLP printing conditions.

2.3. DLP

With a light source operating at a wavelength of $\lambda = 405$ nm, the Admaflex 130 (Admatec Europe BV, Alkmaar, The Netherlands) was employed as a 3D printing machine capable of achieving a resolution of 40 µm on the layer plane. The actual forming tool of the machine is composed of thousands of mirrors that move in accordance with pixels to imprint the picture on the slurry. The Admaflex 130 utilises a rotating foil system for transporting the slurry from the reservoir to the 3D construction area and subsequently to the collecting zone, where a pump mechanism returns the unneeded slurry to the reservoir. The light-reactive slurry is exposed to light at the machine's operating wavelength and, as a result, the entire layer is cured at once due to the presence of multiple mirrors. Gradually, the green body is printed in its totality after being repeatedly cured, layer by layer (Figure 1b).

(a)

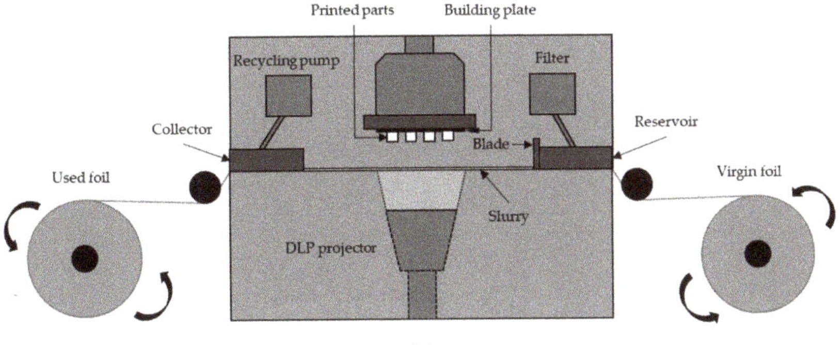

(b)

Figure 1. (**a**) Admaflex 130; (**b**) schematic of the Admaflex 130 DLP process.

A single-layer curing approach was utilised to assess the critical exposure energy and depth of penetration by measuring the cure depth at various exposure energies. The working curve of the ceramic slurry depicts the slurry's curing depth as a function of the light dosage for a particular light source. The vertical polymerisation depth, C_d, may be experimentally stated with Jacob's equation:

$$C_d = D_p \ln(E/E_c), \quad (1)$$

where D_p is the penetration depth—i.e., the distance at which light intensity decreases by e^{-1} [22]; E is the surface exposure energy density; and E_c is the critical surface exposure energy density for the slurry [17], which specifies the minimal energy density necessary for the photopolymerisation of the slurry [50]. Therefore, for higher values of D_p and smaller values of E_c, a lower amount of energy density is required to obtain a certain cure depth C_d. E_c and D_p are believed to be resin-specific metrics that are unaffected by exposure factors [51].

On a layer of photopolymerisable resin, a checkerboard picture was projected at different exposure surface powers for various exposure times. Table 3 illustrates the exposure times and exposure surface powers studied for understanding the curing behaviour.

Table 3. Curing parameters investigated.

Exposure Time [s]			Exposure Surface Power [mW/cm^2]				
1	2	3	3.52	6.69	10.21	13.93	17.24

The cured layers were peeled away from the foil and the thickness of the cured material was measured using a portable micrometre (Mitutoyo Digimatic Micrometer, MDC-25PX, Kawasaki, Japan). Five consecutive measurements were collected, and the average cure depth was determined, together with its standard deviation.

Following the single-layer curing tests, full-factorial design of experiments (DoE) was employed to design an experimental plan and determine the impact of printing settings on the dimensional accuracy. The analysis of variance was utilised to determine the relevant printing factors and their interactions with the process outputs [6]. Exposure power, exposure time, and sliced layer thickness were chosen as the experimental variables. The first variable represents the surface energy density delivered to the slurry for every second of exposure to the light source; the second variable expresses the period of time during which the phopolymerisable slurry is continuously cured under the light source for each layer; the third variable is the height of each individual slice of the model. Three levels were chosen for exposure duration, exposure power, and layer thickness. The exposure power is represented as a fraction of the DLP light source intensity, resulting in irradiance values of 3.52 mW/cm^2 for 50‰, 6.69 mW/cm^2 for 100‰, and 10.21 mW/cm^2 for 150‰. The experimental matrix for the complete factorial experiment is presented in Table 4.

For each set of conditions, the printing of a total of 5 discs (CAD dimensions: diameter = 20 mm; thickness = 1.5 mm) and 5 squared plates (CAD dimensions: side = 15 mm; thickness = 1.5 mm) was attempted. The designed CAD parts were converted to .STL files and sliced to the corresponding layer thickness using the machine's internal slicing software. For each layer, the delay before exposure was set to 15 s to allow the slurry to spread homogeneously on the layer, while the delay after exposure was set to 3 s to allow the layer to stabilise. In addition, the building platform was lifted and lowered each time by 6 mm at a speed of 2.5 mm/s to ensure a complete detachment of the cured layer from the foil and permit safe passage to the sliding foil. Finally, successful or acceptable samples were detached from the building platform and cleaned by using ethanol and compressed air. Minitab 21 statistical software was used to conduct an analysis of variance to determine the statistical relevance of each parameter.

The dimensions of the samples, as well as the geometrical accuracy and the shrinkage, were determined using a combination of a Vernier calliper and an optical microscope Alicona InfiniteFocusG5 plus (Bruker Alicona, Leicestershire, UK). The density of the samples was deducted adopting Archimedes' principle and the optimal printing parameters were selected.

Table 4. DoE factors, levels, and experimental matrix of the DLP experimental campaign.

Factors	Levels		
Exposure power [mW/cm^2]	3.52	6.69	10.21
Exposure time [s]	1	2	3
Layer thickness [μm]	25	50	100
Experiment	**[mW/cm^2]**	**[s]**	**[μm]**
1	3.52	1	25
2	3.52	1	50
3	3.52	1	100
4	3.52	2	25
5	3.52	2	50
6	3.52	2	100
7	3.52	3	25
8	3.52	3	50
9	3.52	3	100
10	6.69	1	25
11	6.69	1	50
12	6.69	1	100
13	6.69	2	25
14	6.69	2	50
15	6.69	2	100
16	6.69	3	25
17	6.69	3	50
18	6.69	3	100
19	10.21	1	25
20	10.21	1	50
21	10.21	1	100
22	10.21	2	25
23	10.21	2	50
24	10.21	2	100
25	10.21	3	25
26	10.21	3	50
27	10.21	3	100

2.4. Thermal Post-Processing

Removing the photopolymerisable resin from the region between the ceramic particles is a vital step in post-processing 3D-printed components. During the debinding process, the organic components of the green body disintegrate into gas molecules, which disperse or pierce the surface of the green body and ultimately escape, creating interspaces between particles inside the brown body. Thermogravimetric analysis (TGA) and a differential scanning calorimetry (DSC) test were performed on an SDT NETZSCH STA with a heating rate of 5 °C/min to determine the mass loss of the DLP-printed green body and the crucial temperature for the binder removal. After printing more samples for the optimal printing parameters, 6 different debinding regimes were explored by varying the heating rate and the holding times, while keeping constant the 3 dwelling temperatures previously set according to the TGA/DSC test. Moreover, a pre-sintering step at 1000 °C was added to promote densification and facilitate the post-debinding sample handling. The debinding cycles were then carried out in a tube furnace (TSH/15/75/450, Elite Thermal Systems Ltd., Market Harborough, UK).

The dimensions and density of the samples were again tested and measured. Finally, the brown parts with the optimal properties were sintered according to the available facilities at 1700 °C for 12 h in a sintering furnace (HTF 17/27, Carbolite Gero, Sheffield, UK) with a heating rate of 10.0 °C/min in an air atmosphere in order to achieve fully dense alumina parts. A schematic summary of the entire experimental campaign is reported in Figure 2.

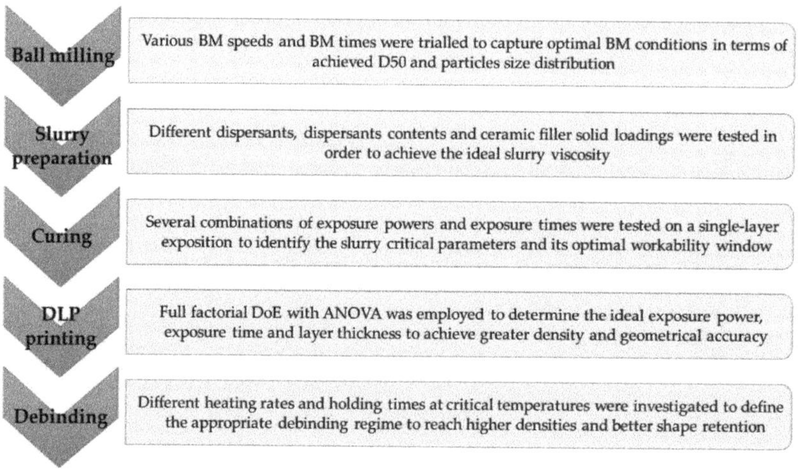

Figure 2. Schematics of the whole experimental campaign.

3. Results and Discussion

3.1. Ball Milling

The various ball-milling (BM) trials resulted in similar results, as the window explored in terms of milling time was chosen according to the existing literature and was quite restricted, as, in this process chain, the ball milling is a preprocessing step that aims to break the agglomerates in the as-received powder rather than significantly changing the powder-size distribution. In addition, longer ball-milling times would increase the slurry preparation time and, therefore, increase the total time duration of the process chain, the completion of which, in the case of additive manufacturing processes, is already relatively time-expensive. Hence, a tight range for ball-milling time was chosen while the ball-milling speed was allowed to change in a wider range, as reported previously in Table 1. Nevertheless, some of the trials, slightly more than other trials, managed to shift the distribution toward a lower average particle size, while breaking the agglomerates in the micron scale and leaving behind mostly sub-micron particles. For brevity, only the most successful trials are reported in Figure 3.

Even though all the results across all the trials conducted were similar, the results corresponding to the lowest level of BM time succeeded in deagglomerating the as-received powder and shifting the D50 0.92 ± 0.02 µm to ~0.75 µm. However, the run with the highest BM speed and, hence, the highest kinetic energy input, managed to achieve a D50 of 0.71 ± 0.01 µm, while also performing better than the other trials, with the same BM time, at reducing the presence of agglomerates. Furthermore, the samples were analysed under the SEM (Figure 4).

Although it is not easy at these magnifications to obtain high resolution, it can be seen that particles seem more deagglomerated as the BM speed is increased, suggesting that at lower BM times it is preferable to have a high kinetic input to thin out the ceramic powders. As a consequence, combining the results obtained from both PSAs and SEM images, the BM speed of 800 rpm and the BM time of 1 h were chosen to continue the investigation.

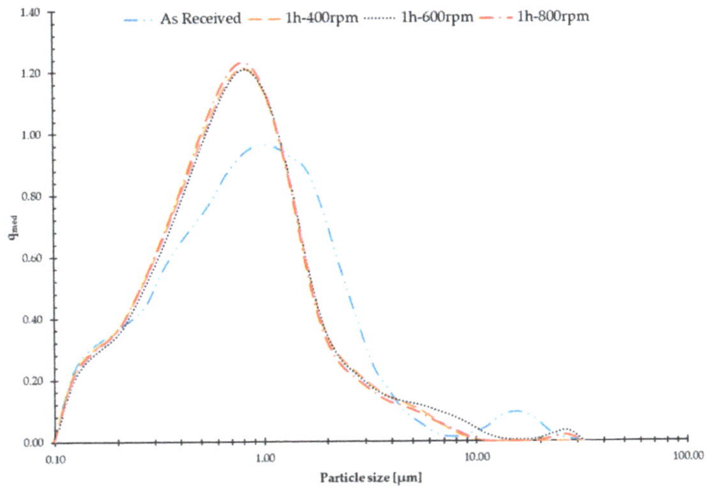

Figure 3. Particle size distribution of the as-received and the most successful BMed powders.

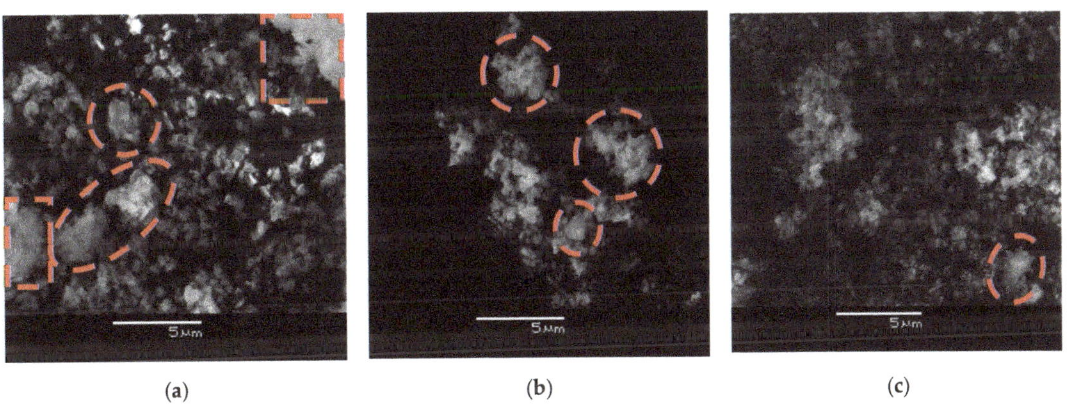

Figure 4. SEM images of BM powders: (**a**) 1 h 400 rpm; (**b**) 1 h 600 rpm; (**c**) 1 h 800 rpm. Visible agglomerates are circled in dashed lines.

3.2. Slurries Characterisation

First, the rheology of the slurries with the chosen different dispersants was tested; the results are reported in Figure 5. It can be noticed that the viscosity of all the prepared slurries decreases at increasing shear rates and their viscosity does not change relevantly from shear rates higher than 20 s^{-1}, which is the suggested behaviour for slurries that have to be used with Admaflex 130. However, they should have a shear-thinning behaviour with a dynamic viscosity below 10 Pa·s for shear rates between 10 and 300 s^{-1} [49]. When utilising a doctor blade gap of 120 μm, typical shear rates throughout the process can reach up to 200 s^{-1} on average. As a matter of fact, the BYK-9076 is not within the workability window previously defined. The slurry with the optimal performance is the one containing BYK-145 as a dispersant, as it possesses the rheology requirements needed to be DLP-printed with Admaflex 130, and both its stability and its lower viscosity, compared to the slurry with BYK-111, allow a better recovering of each layer during the printing process.

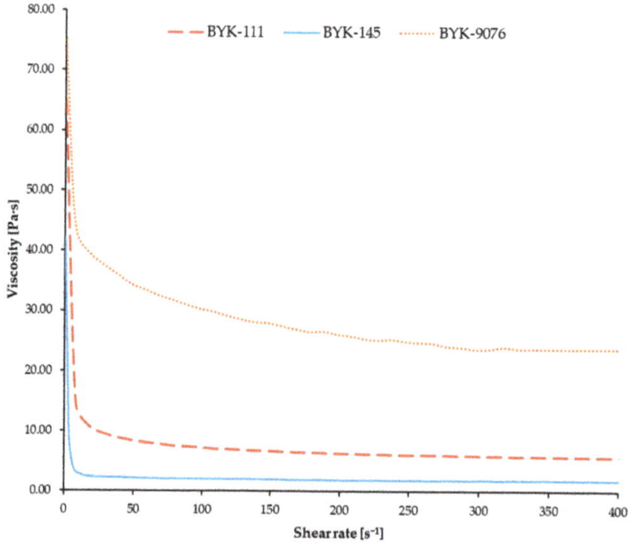

Figure 5. Viscosity vs. shear rate of slurries with different dispersants.

After choosing the optimal dispersant, its weight percentage content was varied to understand the different outcomes in terms of viscosity and to comprehend if there was a saturation value beyond which the dispersant is actually agglomerating rather than dispersing. Viscosity tests were then carried out, and the results are plotted in Figure 6.

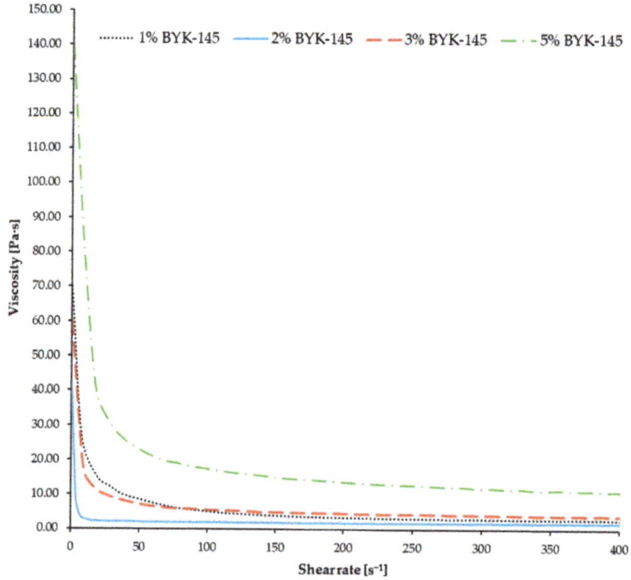

Figure 6. Viscosity vs. shear rate of slurries with BYK-145 at different weight contents.

As can be observed, initially the viscosity of the slurry slightly decreased when increasing the dispersant content from 1 wt.% to 2 wt.%, and it reached a stable value at smaller shear rates, which is the desired behaviour. However, if the content was increased to 3 wt.% or 5 wt.%, the viscosity values in the workability window of Admaflex130 increased. Therefore, the stability of the slurry is degraded if the dispersant concentration is too high [21]. Al_2O_3 particles are not efficiently modified by the dispersant when the concentration is extremely low; however, when its concentration was appropriately raised to 2 wt.%, the particles' surface covering is more efficient, resulting in a more stable network of particles [52]. However, at higher concentrations, there is a surplus of dispersant, which promotes flocculation and, hence, increases the viscosity.

Afterwards, the role of the solid loading was examined; the rheology of the various slurries is reported in Figure 7. As expected, the 70 wt.% solid-loaded slurry exhibited a less viscous behaviour, which resulted in a more liquid slurry being unable to remain within the reservoir during the DLP printing and moved via the tape-casting foil system of Admaflex130 without spilling out. This was due to the extremely low viscosity (<1 Pa·s in the window of interest). On the contrary, the 80 wt.% loading resulted in a more viscous slurry than the 85 wt.% loaded slurry at lower shear rates, and slightly less viscous at higher shear rates. It is possible that the quantity of the dispersant was not optimal for the former, so the particles were flocculating at lower shear rates; however, as higher percentages of solid loading facilitate the densification of the finished samples while reducing the shrinkage and, hence, the insurgence of cracks due to the geometrical modifications, the 85 wt.% was preferred as it also showed better properties without further studies. The 90 wt.% solid-loaded slurry was also prepared, even though it was expected that increasing the packing fraction would have rapidly increased the suspension viscosity [53]. In fact, its consistency was more similar to a powder rather than a slurry, even after long mixing and varying the percentage of dispersant. This was essentially due to the reduction of the average distance between the ceramic particles when the solid loading was increased; as a consequence, a relevant increase in the friction between them resulted in a higher viscosity [54]. Therefore, it was concluded that such a high solid loading was not viable and the 85 wt.% solid-loaded slurry was selected for the continuation of the study.

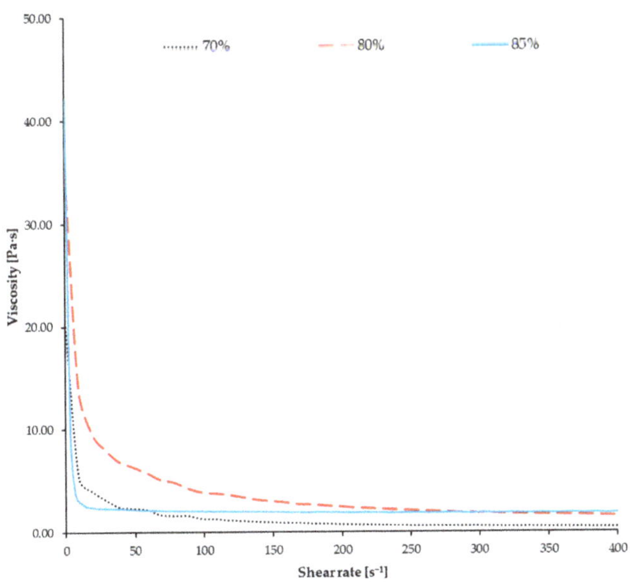

Figure 7. Viscosity vs. shear rate of slurries with different solid loadings.

3.3. *Curing Behaviour and DLP Printing*

The curing behaviour of the selected slurry was analysed by registering several cure depth values at different exposure times and powers, which are reported in Figure 8.

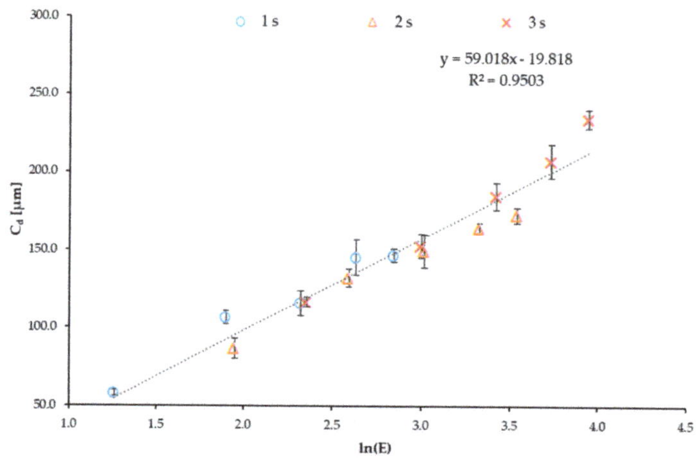

Figure 8. Cure depth values of the slurry at different exposure powers and times.

The values of D_p and E_c can be deducted from the linear regression of the collected data and, in particular, D_p = 59.018 μm and E_c = 1.399 mJ/cm^2. While the latter is compatible with a highly reactive slurry, the former describes a slurry that permits better control of lateral definition, as smaller values of D_p promote higher resolution [55]. In addition, even though all the values were registered while using different exposure parameters, all the collected data points lay acceptably around the trend line obtained from the linear regression with a good correlation coefficient, demonstrating that D_p and E_c are independent of the exposure conditions [51].

Afterwards, the printing trials were started following the schedule established through the DOE. Among all the trials undertaken, the 25 μm layer thickness samples were the most successful in terms of the density achieved and the fidelity of the dimensions to the starting CAD design (CAD diameter = 20 mm; CAD plate side = 15 mm) and, in particular, the ones corresponding to the lowest level of exposure power were the most outstanding, as can be seen in Figure 9 for discs and Figure 10 for plates.

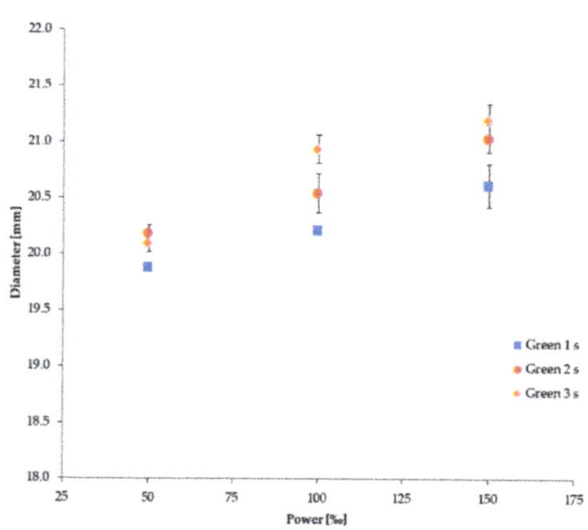

(a)

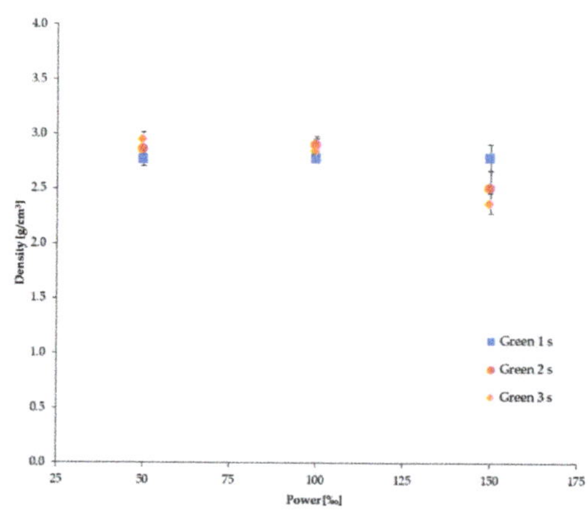

(b)

Figure 9. Diameter dimension (**a**) and measured density (**b**) of the DLPed discs with 25 μm layer thickness.

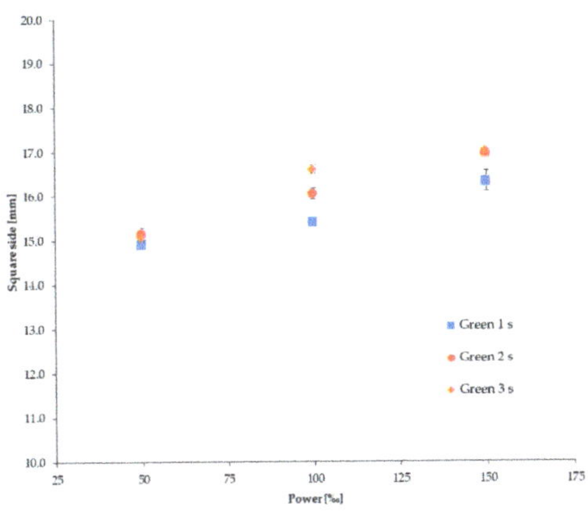

(a)

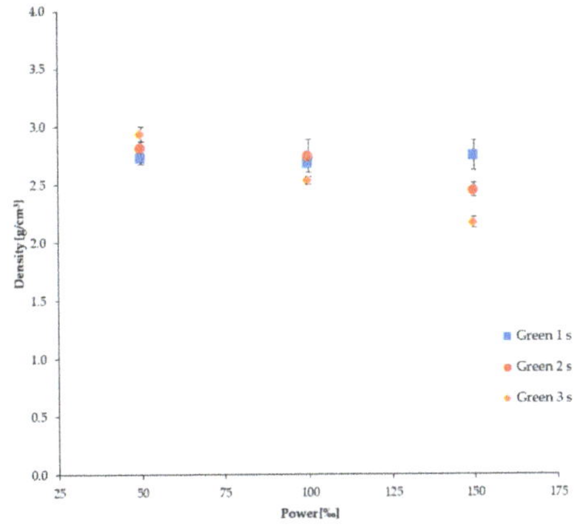

(b)

Figure 10. Side dimension (**a**) and measured density (**b**) of the DLPed plates with 25 μm layer thickness.

The highest setting of power produced inaccurate parts with a high deviation from the corresponding dimension modelled via CAD and, also, a significant standard deviation, as the results were less predictable due to the light scattering phenomenon. Since the energy input was too high, the respective cure depth was the equivalent of a few layers which, hence, were irradiated many times repeatedly. In the case of an exposure time of 3 s and an exposure power of 150‰, the same layer was irradiated almost 10 times, as can be deducted from Figure 8; naturally, this caused relevant overcuring, dimensional inaccuracies, and higher standard deviations (Figure 11). Therefore, parts produced with the lowest exposure power among the samples printed with a layer thickness of 25 μm were considered the best ones (Figure 12) and, specifically, the parts that were exposed for 2 s and 3 s were labelled as the optimal ones, due to their good geometrical accuracy and final green density.

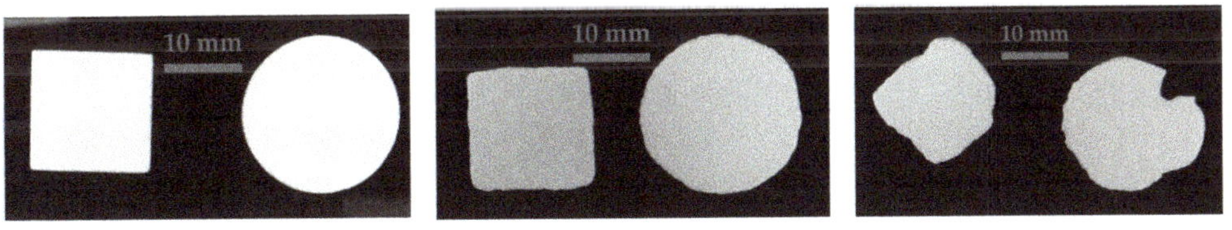

(a) (b) (c)

Figure 11. DLP printed part with 25 μm as layer thickness, 3 s exposure time, and exposure power of (**a**) 50‰; (**b**) 100‰; (**c**) 150‰.

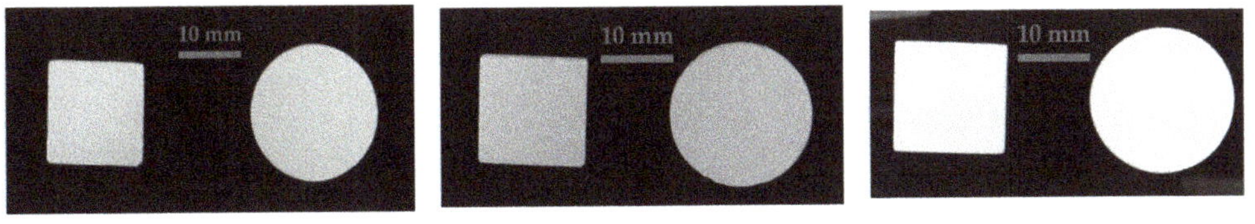

(a) (b) (c)

Figure 12. DLP printed part with 25 μm as layer thickness, 50‰ s exposure time, and exposure time of (**a**) 1 s; (**b**) 2 s; (**c**) 3 s, same as Figure 11a.

For the discs and plates with layer thicknesses of 50 μm and 100 μm, the samples that were tried to be produced at the lowest level of power and exposure time failed, as the energy input was probably not sufficient for guaranteeing enough light penetration to stick two consecutive layers in a stable way. Other conditions were all printed with different outcomes. Since the energy delivered to the layer in the same printing conditions, in terms of exposure time and power, was the same but the volume to which it was delivered was different, the samples with higher layer thickness failed or were mechanically weaker at lower energy input conditions, but survived successfully without heavy distortions or deviations at higher energetic conditions. However, due to the fact that for a specified thickness, the number of layers diminishes with the increase of the layer thickness, the 50 μm and, in particular, the 100 μm samples lacked a good interlayer cohesion, resulting in samples that had to be handled carefully and were slightly bendable. On the other hand, the 25 μm samples were more compacted and rigid and, thus, achieved a greater density, which was the main objective for producing, in the end, translucent parts. It was interesting to notice that, in general, across all the samples, an increase in the exposure time level, rather than an increase in the exposure power level when the same energy dose was delivered, led to better results, suggesting that the management of the exposure time is more significant than the management of exposure power. In fact, the slower energy delivery to the slurry when longer exposure times and lower exposure powers are adopted allows the layers of the green parts to be cured more gradually from the uncured slurry surface directly exposed to the DLP projector up to the previously cured layers within the curing depth, ensuring a more stable printing, better interlayer bonding, and increasing geometric accuracy. On the contrary, conveying the same energy dose by utilising higher exposure power levels and lower exposure times is more similar to an impulsive input of energy, which is less controllable and less accurate, resulting in a horizontal overcuring on the layer plane. Therefore, combinations of lower exposure powers and longer exposure times should be preferred when the optimal energy dose is already determined. It is very important to select the optimal set of parameters, as final sintered parts properties are dependent on the curing process during DLP printing, since the polymerisation grade of the green parts directly affects the bonded geometry of the unsintered alumina [56]. In addition, the high solid loading of small-sized ceramic particles ensures that the achievable final density is higher, as the capability to reach a greater sustainable packing factor allows a better densification [57].

Table 5 reports the significance values for the parameters investigated and their interactions, as obtained from Minitab 21, when density is imposed as the response to analyse, while Figures 13 and 14 illustrate the factorial plots for the density.

Following the previous findings that confirmed that the significant process parameters consisted of the exposure time and its interaction with the exposure power, since their significance value was $p < 0.05$, the printing parameters of 25 μm as layer thickness, 2 s as exposure time, and 50‰ as exposure power were considered the optimal parameters and were selected to print more discs for the following debinding experiments. It must be noted that the low values of mean density corresponding to the lowest levels of exposure power and time were mainly due to the failed trials for the 50 μm and 100 μm layer thicknesses.

Table 5. *p*-values of analysed factors and their interactions.

Factor	*p*-Value
Exposure power	0.109
Exposure time	0.027
Layer thickness	0.066
Exposure power – Exposure time	0.023
Exposure power – Layer thickness	0.269
Exposure time – Layer thickness	0.246

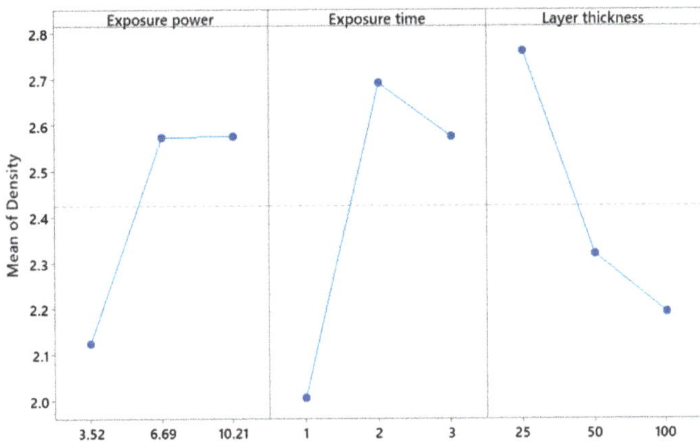

Figure 13. Effect of the process parameters on the mean of the density.

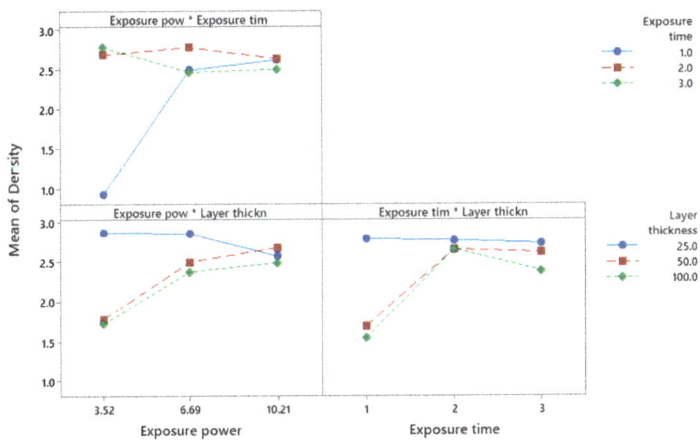

Figure 14. Effect of the process parameters interactions on the mean of the density.

3.4. Thermal Post-Processing

The TGA–DSC study of the green body provided data on the breakdown behaviour of the organic materials in relation to the fluctuating temperature (Figure 15). At around 375 °C and 480 °C, two significant exothermic peaks were recorded and, as seen by the TG curve, the total mass loss was 15.29%, indicating the complete burnout of the organic components. The first mass loss began to occur when the temperature was between 150 °C and 250 °C. This weight loss can be ascribed to the evaporation of adsorbed water and residual cleaning solution. The second mass loss took place between 270 °C and 380 °C, while the third and last significant loss occurred between 390 °C and 550 °C, beyond which the mass loss was negligible. Since the starting ceramic powder was α-Al_2O_3, from the DSC curve it can be deduced that the thermal debinding reaction was an exothermic process with two major peaks, the former attributed to the oxidative decomposition of polymer chains and the latter ascribed to the thermal decomposition of the crosslinked polymer network [58]. The debinding regime was established, based on these findings. Figure 16 illustrates the temperature settings for the debinding procedure in detail. The temperature was increased from room temperature to 175 degrees Celsius and allowed to dwell at this initial temperature setpoint. The temperature was then raised to 375 degrees Celsius and sustained for another dwelling-time period in order to degrade the organics slowly and, as a result, alleviate the thermal strains on the ceramic components. Afterwards, the system was heated to 480 degrees Celsius and kept at this temperature. Finally, the temperature was steadily increased to 1000 degrees Celsius to eradicate any organic residue and to pre-sinter the sample, which was held for another dwelling and, then, left to cool down to room temperature. Throughout the whole debinding procedure, the rate of heating was regulated according to the ongoing scheduled trial. Figure 16 illustrates the debinding regimes investigated.

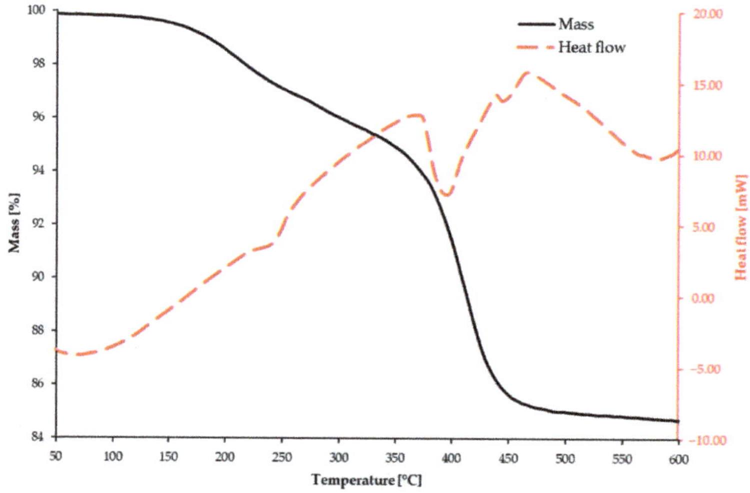

Figure 15. TG–DSC curves of the DLPed Al_2O_3 sample.

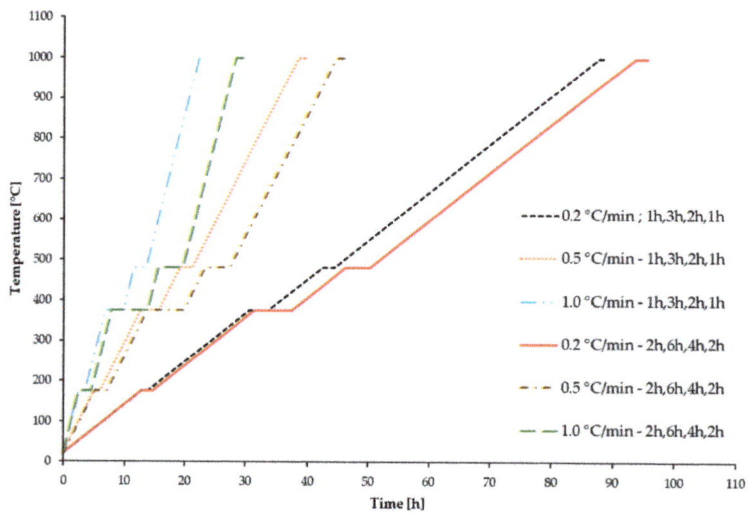

Figure 16. Debinding regimes explored.

After undergoing debinding, the samples were again tested and measured to observe the geometrical dimensions and density evolution. The outcomes of these analyses are reported in Figure 17.

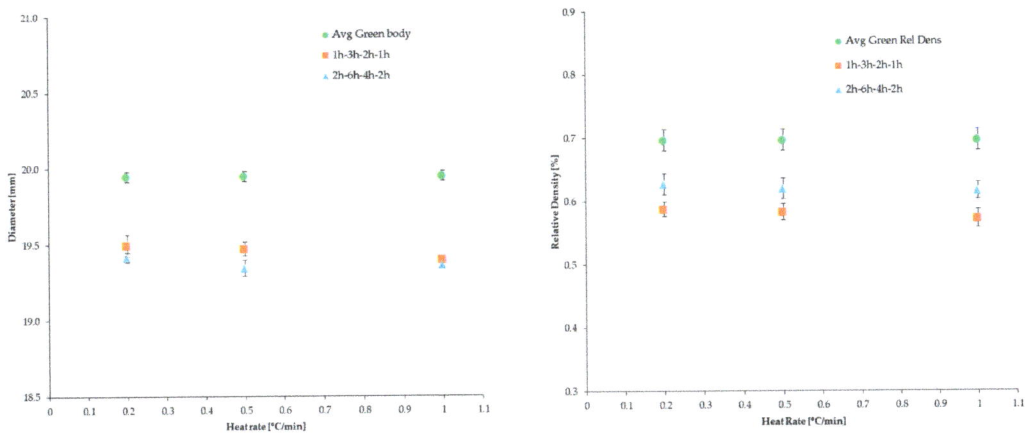

(a) (b)

Figure 17. Diameter dimension (**a**) and measured density (**b**) of the debound discs.

It was observed that at lower heating rates and longer dwelling times, the samples were able to reach higher densities, as the gradual removal of the binder over a longer time contributed to the reduction of residual pores [59] and, hence, to the reaching of higher brown densities. In fact, small heating rates when transitioning between peak dwelling temperatures provided more time for the organics to escape the ceramic structure, instead of being abruptly extracted and leaving behind cracks and defects. Moreover, longer holding times at significant dwelling temperatures ensured that all the corresponding organics exiting the sample were completely expelled. On the contrary, the highest heat rate of 1.0 °C/min, while allowing the samples to reach comparable densities to the ones obtained from the cycles at 0.2 °C/min, resulted in a lower diameter dimension, due to the rapid binder extraction, which caused loss of superficial debris and, hence, geometrical inaccuracy (Figure 18).

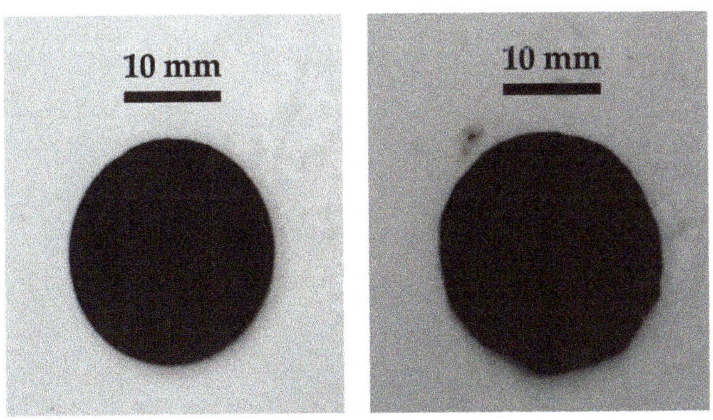

(a) (b)

Figure 18. (**a**) Sample debound at 0.2 °C/min for the longer set of holding times; (**b**) sample debound at 1.0 °C/min for the longer set of holding times.

Following the previous results, the debinding cycle with a heat rate of 0.2 °C/min and dwelling times of 2 h, 6 h, 4 h, and 2 h was chosen to conclude this investigation. The samples were then prepared, following the process recipe established in this investigation, and finally heated up in an air sintering furnace at a heating rate of 10 °C/min until the temperature of 1700 °C was reached, then held in this condition for 12 h, after which they spontaneously cooled down inside the sintering furnace chamber. Another batch of samples was prepared following this process chain until the end, with the exception of adding MgO as a sintering aid in the starting slurry in a percentage of ~1000 ppm to promote densification and help with the management of the grain growth. Both samples exceeded the relative density of 98.5% and, in particular, the sample with MgO registered a relative density slightly higher than 99.0%, thereby starting to demonstrate a translucent behaviour (Figure 19).

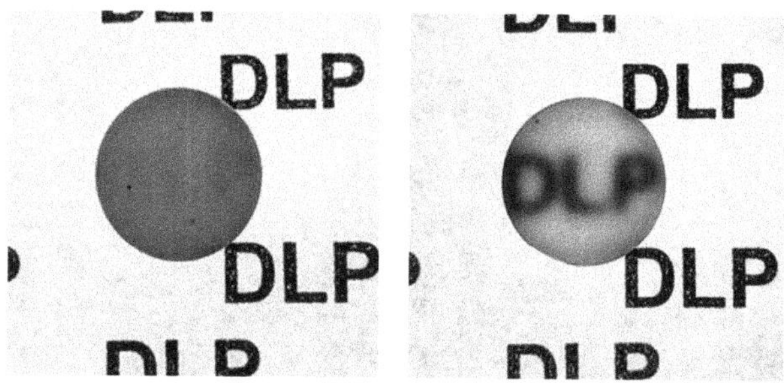

(a) (b)

Figure 19. (**a**) Sample prepared following the process recipe herein described; (**b**) sample prepared following the process recipe herein described with the addition of MgO as a sintering aid.

4. Conclusions

A systematic method for determining the optimal components and process parameters for developing an efficient slurry in-house from the ball-milling parameters selection down to the debinding regimes to follow was extensively presented. This method can be applied to tailor the slurry properties to the desired values and develop specific slurries for particular scopes, such as, in this case, producing final parts that exhibit translucency, by working on constantly maintaining high values of relative densities. The main findings of this study can be summarised as follows.

- Within the workability window explored, dry ball milling with low milling time and high milling speed (1 h and 800 rpm) was the most effective in reducing agglomerates to sharpen the as-received powder size distribution.
- Among the investigated various dispersants and their concentrations, the most appropriate dispersant to use with the specific slurry prepared in this study was BYK-145 with a weight percentage content of 2% to the ceramic powder content.
- Different solid loadings were examined and ~85 wt.% of ceramic filler resulted in the optimal choice, reducing the risk of heavy distortions and excessive thermal stress while guaranteeing high densities at all stages.
- The curing behaviour of the prepared slurry was analysed and the conducted full-factorial DoE highlighted that exposure time and its interaction with the exposure power were the most significant parameters in the DLP printing of the developed slurry. Optimal printing parameters were 2 s as exposure time and 50‰ as exposure power when the layer thickness was set to 25 μm.
- Among all the debinding trials investigated, lower heating rates (0.2 °C/min) and longer dwelling times delivered a more stable sample with higher relative density and geometrical dimensions retention.
- Finally, samples were manufactured according to the process schedule established and sintered at 1700 °C for 12 h. The ones with the addition of MgO (1000 ppm) as a sintering aid started to demonstrate a translucent behaviour.

Future development of this work should be aimed at achieving more dense and translucent samples by utilising a high-performance sintering furnace able to work under vacuum at high temperatures or a microwave sintering furnace. In addition, a study on the effect of sintering parameters, such as heating rate and dwelling time as well as dwelling temperature, may provide useful information on the optimal sintering cycle to follow and result in decreased light refraction. In fact, to obtain higher translucency, the final relative density of the parts must be increased further over 99.5% while controlling and tailoring the grain growth to maximise light transmittance. This could be achieved by sintering the parts under vacuum at high temperatures (>1500 °C) in order to minimise air encapsulation and, hence, residual porosity, while improving the densification process.

Author Contributions: Conceptualization, methodology, investigation, formal analysis, data curation, and writing—original draft preparation, visualization, M.D.L.; investigation and writing—review and editing, C.S.; supervision and writing—review and editing, U.M.A. and K.E. All authors have read and agreed to the published version of the manuscript.

Funding: The authors would like to thank the Manufacturing Technology Centre (MTC) for the financial support of Michele De Lisi's PhD project (Manufacturing Technology Centre, United Kingdom: 881440).

Data Availability Statement: All data are provided within the article and the provided references.

Acknowledgments: M.D.L. thanks the University of Birmingham and the Manufacturing Technology Centre (MTC) for the award of a PhD scholarship.

Conflicts of Interest: The authors declare no conflict of interest. The funders had no role in the design of the study; in the collection, analyses, or interpretation of data; in the writing of the manuscript; or in the decision to publish the results.

References

1. Bikas, H.; Stavropoulos, P.; Chryssolouris, G. Additive manufacturing methods and modeling approaches: A critical review. *Int. J. Adv. Manuf. Technol.* **2016**, *83*, 389–405. [CrossRef]
2. Chen, Z.; Li, Z.; Li, J.; Liu, C.; Lao, C.; Fu, Y.; Liu, C.; Li, Y.; Wang, P.; He, Y. 3D printing of ceramics: A review. *J. Eur. Ceram. Soc.* **2019**, *39*, 661–687. [CrossRef]
3. Yang, L.; Miyanaji, H. Ceramic additive manufacturing: A review of current status and challenges. In Proceedings of the 28th Annual International Solid Freeform Fabrications Symposium—An Additive Manufacturing Conference, SFF 2017, Austin, TX, USA, 7–9 August 2017. Available online: https://hdl.handle.net/2152/89871 (accessed on 2 November 2021).
4. Zocca, A.; Colombo, P.; Gomes, C.M.; Günster, J. Additive manufacturing of ceramics: Issues, potentialities, and opportunities. *J. Am. Ceram. Soc.* **2015**, *98*, 1983–2001. [CrossRef]
5. Jasveer, S.; Jianbin, X. Comparison of different types of 3D printing technologies. *Int. J. Sci. Res. Publ.* **2018**, *8*, 1–9. [CrossRef]
6. Farzadi, A.; Solati-Hashjin, M.; Asadi-Eydivand, M.; Osman, N.A.A. Effect of layer thickness and printing orientation on mechanical properties and dimensional accuracy of 3D printed porous samples for bone tissue engineering. *PLoS ONE* **2014**, *9*, e108252. [CrossRef] [PubMed]
7. Tan, C.; Li, S.; Essa, K.; Jamshidi, P.; Zhou, K.; Ma, W.; Attallah, M.M. Laser Powder Bed Fusion of Ti-rich TiNi lattice structures: Process optimisation, geometrical integrity, and phase transformations. *Int. J. Mach. Tools Manuf.* **2019**, *141*, 19–29. [CrossRef]
8. Hassanin, H.; Zweiri, Y.; Finet, L.; Essa, K.; Qiu, C.; Attallah, M. Laser Powder Bed Fusion of Ti-6Al-2Sn-4Zr-6Mo Alloy and Properties Prediction Using Deep Learning Approaches. *Materials* **2021**, *14*, 2056. [CrossRef]
9. Bittredge, O.; Hassanin, H.; El-Sayed, M.A.; Eldessouky, H.M.; Alsaleh, N.A.; Alrasheedi, N.H.; Essa, K.; Ahmadein, M. Fabrication and Optimisation of Ti-6Al-4V Lattice-Structured Total Shoulder Implants Using Laser Additive Manufacturing. *Materials* **2022**, *15*, 3095. [CrossRef]
10. El-Sayed, M.A.; Essa, K.; Ghazy, M.; Hassanin, H. Design optimization of additively manufactured titanium lattice structures for biomedical implants. *Int. J. Adv. Manuf. Technol.* **2020**, *110*, 2257–2268. [CrossRef]
11. Essa, K.; Hassanin, H.; Attallah, M.M.; Adkins, N.J.; Musker, A.J.; Roberts, G.T.; Tenev, N.; Smith, M. Development and testing of an additively manufactured monolithic catalyst bed for HTP thruster applications. *Appl. Catal. A Gen.* **2017**, *542*, 125–135. [CrossRef]
12. Fu, X.; Zou, B.; Xing, H.; Li, L.; Li, Y.; Wang, X. Effect of printing strategies on forming accuracy and mechanical properties of ZrO2 parts fabricated by SLA technology. *Ceram. Int.* **2019**, *45*, 17630–17637. [CrossRef]
13. Ozog, P.; Blugan, G.; Kata, D.; Graule, T. Influence of the printing parameters on the quality of alumina ceramics shaped by UV-LCM technology. *J. Ceram. Sci. Technol.* **2019**, *10*, 1–10. [CrossRef]
14. Sun, C.; Zhang, X. The influences of the material properties on ceramic microstereolithography. *Sens. Actuators A.* **2002**, *101*, 364–370. [CrossRef]

15. Gentry, S.P.; Halloran, J.W. Depth and width of cured lines in photopolymerizable ceramic suspensions. *J. Eur. Ceram. Soc.* **2013**, *33*, 1981–1988. [CrossRef]
16. Hu, K.; Wei, Y.; Lu, Z.; Wan, L.; Li, P. Design of a shaping system for stereolithography with high solid loading ceramic suspensions. *3D Print. Addit. Manuf.* **2018**, *5*, 311–318. [CrossRef]
17. Komissarenko, D.A.; Sokolov, P.S.; Evstigneeva, A.D.; Shmeleva, I.A.; Dosovitsky, A.E. Rheological and curing behavior of acrylate-based suspensions for the DLP 3D printing of complex zirconia parts. *Materials* **2018**, *11*, 2350. [CrossRef]
18. Varghese, G.; Moral, M.; Castro-García, M.; López-López, J.J.; Marín-Rueda, J.R.; Yagüe-Alcaraz, V.; Hernández-Afonso, L.; Ruiz-Morales, J.C.; Canales-Vázquez, J. Fabrication and characterisation of ceramics via low-cost DLP 3D printing. *Bol. Soc. Esp. Ceram. Vidrio* **2018**, *57*, 9–18. [CrossRef]
19. Sokolov, P.S.; Komissarenko, D.A.; Shmeleva, I.A.; Slyusar, I.V.; Dosovitskiy, G.A.; Evdokimov, P.V.; Putlyaev, V.I.; Dosovitskiy, A.E. Suspensions on the basis of stabilised zirconium oxide for three-dimensional printing. *IOP Conf. Ser. Mater. Sci. Eng.* **2018**, *347*, 012012. [CrossRef]
20. Mitteramskogler, G.; Gmeiner, R.; Felzmann, R.; Gruber, S.; Hofstetter, C.; Stampfl, J.; Ebert, J.; Wachter, W.; Laubersheimer, J. Light curing strategies for lithography-based additive manufacturing of customized ceramics. *Addit. Manuf.* **2014**, *1*, 110–118. [CrossRef]
21. Zhang, K.; Xie, C.; Wang, G.; He, R.; Ding, G.; Wang, M.; Dai, D.; Fang, D. High solid loading, low viscosity photosensitive Al_2O_3 slurry for stereolithography based additive manufacturing. *Ceram. Int.* **2019**, *45*, 203–208. [CrossRef]
22. Abouliatim, Y.; Chartier, T.; Abelard, P.; Chaput, C.; Delage, C. Optical characterization of stereolithography alumina suspensions using the Kubelka-Munk model. *J. Eur. Ceram. Soc.* **2009**, *29*, 919–924. [CrossRef]
23. Sun, J.; Binner, J.; Bai, J. Effect of surface treatment on the dispersion of nano zirconia particles in non-aqueous suspensions for stereolithography. *J. Eur. Ceram. Soc.* **2019**, *39*, 1660–1667. [CrossRef]
24. Wang, J.C.; Dommati, H. Fabrication of zirconia ceramic parts by using solvent based slurry stereolithography and sintering. *Int. J. Adv. Manuf. Technol.* **2018**, *98*, 1537–1546. [CrossRef]
25. Griffith, M.L.; Halloran, J. Ultraviolet curing of highly loaded ceramic suspensions for stereolithography of ceramics. In Proceedings of the International Solid Free Fabrication Symposium, Austin, Texas, USA, 8–10 August 1994; pp. 396–403. [CrossRef]
26. Song, S.Y.; Park, M.S.; Lee, D.; Lee, J.W.; Yun, J.S. Optimization and characterization of high-viscosity ZrO_2 ceramic nanocomposite resins for supportless stereolithography. *Mater. Des.* **2019**, *180*, 107960. [CrossRef]
27. Shuai, X.; Zeng, Y.; Li, P.; Chen, J. Fabrication of fine and complex lattice structure Al_2O_3 ceramic by digital light processing 3D printing technology. *J. Mater. Sci.* **2020**, *55*, 6771–6782. [CrossRef]
28. Guo, J.; Zeng, Y.; Li, P.; Chen, J. Fine lattice structural titanium dioxide ceramic produced by DLP 3D printing. *Ceram. Int.* **2019**, *45*, 23007–23012. [CrossRef]
29. Wang, K.; Qiu, M.; Jiao, C.; Gu, J.; Xie, D.; Wang, C.; Tang, X.; Wei, Z.; Shen, L. Study on defect-free debinding green body of ceramic formed by DLP technology. *Ceram. Int.* **2020**, *46*, 2438–2446. [CrossRef]
30. De Lisi, M.; Kovacev, N.; Attia, U.M.; Essa, K. Numerical Simulation of Sintering of DLP Printed Alumina Ceramics. *Aerospace* **2022**, *9*, 336. [CrossRef]
31. Hassanin, H.; Essa, K.; Elshaer, A.; Imbaby, M.; El-Mongy, H.H.; El-Sayed, T.A. Micro-fabrication of ceramics: Additive manufacturing and conventional technologies. *J. Adv. Ceram.* **2021**, *10*, 1–27. [CrossRef]
32. Kovacev, N.; Li, S.; Zeraati-Rezaei, S.; Hemida, H.; Tsolakis, A.; Essa, K. Effects of the internal structures of monolith ceramic substrates on thermal and hydraulic properties: Additive manufacturing, numerical modelling and experimental testing. *Int. J. Adv. Manuf. Technol.* **2021**, *112*, 1115–1132. [CrossRef]
33. Kovacev, N.; Li, S.; Li, W.; Zeraati-Rezaei, S.; Tsolakis, A.; Essa, K. Additive Manufacturing of Novel Hybrid Monolithic Ceramic Substrates. *Aerospace* **2022**, *9*, 255. [CrossRef]
34. Kovacev, N.; Doustdar, O.; Li, S.; Tsolakis, A.; Herreros, J.M.; Essa, K. The synergy between substrate architecture of 3D-printed catalytic converters and hydrogen for low-temperature aftertreatment systems. *Chem. Eng. Sci.* **2023**, *269*, 118490. [CrossRef]
35. Klein, C.A. Figures of merit for high-energy laser-window materials: Thermal lensing and thermal stresses. *Laser-Induc. Damage Opt. Mater. 2006* **2007**, *6403*, 40308. [CrossRef]
36. Moynihan, C.T.; Loehr, S.R. Chemical durability of fluoride glasses. *Mater. Sci. Forum.* **1988**, *32–33*, 243–253. [CrossRef]
37. Bayya, S.; Sanghera, J.S.; Aggarwal, I.D.; Chin, G. Vis-IR transmitting window materials. *SPIE* **2005**, *5786*, 262–271. [CrossRef]
38. Goldman, L.M.; Hartnett, T.M.; Wahl, J.M.; Ondercin, R.J.; Olson, K. Recent advances in aluminum oxynitride (ALON) optical ceramic. In Proceedings of the SPIE Window and Dome Technologies and Materials VII, Orlando, FL, USA, 7 September 2001; Volume 4375, pp. 71–78. [CrossRef]
39. Beyer, R.A.; Kerwien, H. Evaluation of ALON for cannon window application. In Proceedings of the SPIE Window and dome technologies and materials VI, Orlando, FL, USA, 26 July 1999; Volume 3705, pp. 113–118. [CrossRef]
40. Goldman, L.M.; Twedt, R.; Balasubramanian, S.; Sastri, S. ALON optical ceramic transparencies for window, dome and transparent armor applications. In Proceedings of the SPIE Window and dome technologies and materials XII, Orlando, FL, USA, 20 May 2011; Volume 8016. [CrossRef]

41. Askinazi, J.; Wientzen, R.V.; Khattak, C.P. Development of large aperture, monolithic sapphire optical windows. In Proceedings of the SPIE Window and Dome Technologies and Materials VII, Orlando, FL, USA, 7 September 2001; Volume 4375, pp. 1–11. [CrossRef]
42. Patterson, M.C.L.; DiGiovanni, A.; Roy, D.W.; Gilde, G. Spinel armor—Clearly the way to go. In *Ceramic Engineering and Science Proceedings: 27th Annual Cocoa Beach Conference on Advanced Ceramics and Composites: A*; Wiley-Blackwell on behalf of the American Ceramic Society: Columbus, OH, USA, 2003; Volume 24, pp. 441–446. [CrossRef]
43. Patterson, M.C.L.; DiGiovanni, A.A.; Fehrenbacher, L.; Roy, D.W. Spinel: Gaining momentum in optical applications. In Proceedings of the SPIE Window and Dome Technologies and Materials VIII, Orlando, FL, USA, 26 September 2003; Volume 5078, pp. 71–79. [CrossRef]
44. Jones, I.K.; Seeley, Z.M.; Cherepy, N.J.; Duoss, E.B.; Payne, S.A. Direct ink write fabrication of transparent ceramic gain media. *Opt. Mater.* **2018**, *75*, 19–25. [CrossRef]
45. Seeley, Z.; Yee, T.; Cherepy, N.; Drobshoff, A.; Herrera, O.; Ryerson, R.; Payne, S.A. 3D printed transparent ceramic YAG laser rods: Matching the core-related refractive index. *Opt. Mater.* **2020**, *107*, 110121. [CrossRef]
46. Zhang, G.; Carloni, D.; Wu, Y. 3D printing of transparent YAG ceramics using copolymer-assisted slurry. *Ceram. Int.* **2020**, *46*, 17130–17134. [CrossRef]
47. Hostaša, J.; Schwentenwein, M.; Toci, G.; Esposito, L.; Brouczek, D.; Piancastelli, A.; Pirri, A.; Patrizi, B.; Vannini, M.; Biasini, V. Transparent laser ceramics by stereolithography. *Scr. Mater.* **2020**, *187*, 194–196. [CrossRef]
48. Cooperstein, I.; Indukuri, S.R.K.C.; Bouketov, A.; Levy, U.; Magdassi, S. 3D printing of micrometer-sized transparent ceramics with on-demand optical-gain properties. *Adv. Mater.* **2020**, *32*, 2001675. [CrossRef]
49. Gonzalez, P.; Schwarzer, E.; Scheithauer, U.; Kooijmans, N.; Moritz, T. Additive manufacturing of functionally graded ceramic materials by stereolithography. *J. Vis. Exp.* **2019**, *2019*, 1–8. [CrossRef]
50. Griffith, M.L.; Halloran, J.W. Scattering of ultraviolet radiation in turbid suspensions. *J. Appl. Phys.* **1997**, *81*, 2538–2546. [CrossRef]
51. Bennett, J. Measuring UV curing parameters of commercial photopolymers used in additive manufacturing. *Addit. Manuf.* **2017**, *18*, 203–212. [CrossRef] [PubMed]
52. Tomasik, P.; Schilling, C.H.; Jankowiak, R.; Kim, J.C. The role of organic dispersants in aqueous alumina suspensions. *J. Eur. Ceram. Soc.* **2003**, *23*, 913–919. [CrossRef]
53. Wang, D.; Chen, T.; Zeng, Y.; Chen, X.; Xing, W.; Fan, Y.; Qiao, X. Optimization of UV-curable alumina suspension for digital light processing of ceramic membranes. *J. Membr. Sci.* **2022**, *643*, 120066. [CrossRef]
54. Chèvremont, W.; Chareyre, B.; Bodiguel, H. Quantitative study of the rheology of frictional suspensions: Influence of friction coefficient in a large range of viscous numbers. *Phys. Rev. Fluids* **2019**, *4*, 064302. [CrossRef]
55. Delhote, N.; Bila, S.; Baillargeat, D.; Chartier, T.; Verdeyme, S. Advanced design and fabrication of microwave components based on shape optimization and 3D ceramic stereolithography process. In *Advances in Ceramics—Synthesis and Characterization, Processing and Specific Applications*; Intechopen: London, UK, 2011; Volume 11, pp. 243–276. [CrossRef]
56. Cramer, C.L.; Wilt, J.K.; Campbell, Q.A.; Han, L.; Saito, T.; Nelson, A.T. Accuracy of stereolithography printed alumina with digital light processing. *Open Ceram.* **2021**, *8*, 100194. [CrossRef]
57. Xu, X.; Zhou, S.; Wu, J.; Liu, S.; Ma, S.; Cheng, T. Study of alumina ceramic parts fabricated via DLP stereolithography using powders with different sizes and morphologies. *Int. J. Appl. Ceram. Technol.* **2023**, *20*, 1167–1193. [CrossRef]
58. Wang, L.; Liu, X.; Wang, G.; Tang, W.; Li, S.; Duan, W.; Dou, R. Partially Stabilized Zirconia Moulds Fabricated by Stereolithographic Additive Manufacturing via Digital Light Processing. *Mater. Sci. Eng. A* **2020**, *770*, 138537. [CrossRef]
59. Sim, J.H.; Koo, B.K.; Jung, M.; Kim, D.S. Study on Debinding and Sintering Processes for Ceramics Fabricated Using Digital Light Processing (DLP) 3D Printing. *Processes* **2022**, *10*, 2467. [CrossRef]

machines

MDPI

Article

Effects of Tool Edge Geometry and Cutting Conditions on the Performance Indicators in Dry Turning AISI 1045 Steel

Adel T. Abbas [1,*], Magdy M. El Rayes [1], Abdulhamid A. Al-Abduljabbar [1], Adham E. Ragab [2], Faycal Benyahia [1] and Ahmed Elkaseer [3,4,5,*]

1 Department of Mechanical Engineering, College of Engineering, King Saud University, P.O. Box 800, Riyadh 11421, Saudi Arabia
2 Department of Industrial Engineering, College of Engineering, King Saud University, P.O. Box 800, Riyadh 11421, Saudi Arabia
3 Department of Production Engineering and Mechanical Design, Faculty of Engineering, Port Said University, Port Fouad 42526, Egypt
4 Department of Mechanical Engineering, Faculty of Engineering, The British University in Egypt (BUE), El-Sherouk City 11837, Egypt
5 Institute for Automation and Applied Informatics, Karlsruhe Institute of Technology, 76344 Eggenstein-Leopoldshafen, Germany
* Correspondence: aabbas@ksu.edu.sa (A.T.A.); ahmed.elkaseer@kit.edu (A.E.)

Abstract: This article presents an experimental investigation and statistical analysis of the effects of cutting conditions on the machining performance of AISI 1045 steel using a wiper-shaped insert. Experimental findings are used to compare the machining performance obtained using wiper inserts with those obtained using conventional round-nose inserts as recently reported in the literature. In addition, the effects of process conditions, namely cutting speed, feed rate, and depth of cut, are analyzed in order to obtain optimum conditions for both types of inserts. The goal is to achieve the optimal machining outcomes: minimum surface roughness, resultant cutting force, and cutting temperature, but maximum material removal rate. A full factorial design was followed to conduct the experimental trials, while ANOVA was utilized to estimate the effect of each factor on the process responses. A desirability function optimization tool was used to optimize the studied responses. The results reveal that the optimum material removal rate for wiper-shaped inserts is 67% more than that of conventional inserts, while maintaining a 0.7 μm surface roughness value. The superior results obtained with wiper-shaped inserts allow wiper tools to use higher feed rates, resulting in larger material removal rates while obtaining the same surface quality.

Keywords: AISI 1045; dry turning; cutting forces; cutting temperature; surface roughness; wiper-shaped insert; conventional round-nose insert

1. Introduction

Machining has been considered a prime choice for processing a wide range of engineering materials. This can mainly be explained through its high ability to produce complex features with a tight tolerance and high accuracy [1]. However, machining is considered a complex multiphysics process entailing mechanical, thermal, and even chemical regimes [2]. The effect of the process conditions on different performance indicators of the process, such as material removal rate, surface quality, life of cutting tool, cutting forces, and consumed thermal energy, has been widely examined by a number of researchers [1,2].

To examine the effects of nanofluid-based advanced cooling on the performance of machining of AISI1045, Abbas et al. conducted a comparative study of this type of cooling when compared with conventional dry and flood cooling strategies, considering aspects of the product, i.e., surface quality, and aspects of the process, such as energy consumption, in machining AISI1045 steel. Based on the developed mathematical models for the machining responses, the results showed that nanofluid minimum quantity lubrication, with an overall weighted sustainability index of 0.7, exhibited the most sustainable performance and produced the lowest surface roughness and energy consumption. The optimum results (with desirability of 0.9050) were speed of cut = 116 m/min, cutting depth = 0.25 mm, and feeding rate = 0.06 mm/rev. Moreover, lowering the feeding rate was suggested to improve surface quality. To lower power consumption, lowering control factors were recommended [3].

Brown et al. [4] compared and de-convoluted the effects of surface quality resulting from the relations between geometric parameters during the turning process. They concluded that employing large stones for cutting edge geometries reduces the roughness of the machined surface for cases of high values of kinematic roughness. On the other hand, for cases with low values of predicted kinematic roughness, using large stones results in increases in surface roughness, since ploughing occurs in this case. Moreover, instability and side flow dominate conditions and result in larger surface roughness at low kinematic roughness. Process stability and smaller tip size produce better surface roughness at higher kinematic roughness parameters.

Khidhir et al. experimentally evaluated the resultant surface finish and wear of the tool tip when turning nickel-based Hastelloy C-276 under different turning conditions and with two different ceramic cutting inserts. The authors concluded that the interaction between cutting depth and cutting speed produced a built-up edge at low to medium speeds, which affected surface roughness and tool wear. Additionally, examination of SEM images demonstrated that the wear of the nose radius was responsible for the generated high surface roughness values. On the other hand, conventional round inserts resulted in improved surface roughness with a reduction in depth of cut and higher cutting speed [5].

Rodrigues et al. [6] studied the machinability of hardened ASTM H13 steels (50 HRC) at both mild speeds and high speeds of cutting (also known as HSC). Dry tests were conducted on seven different geometries of chip breakers of carbides coated with titanium nitride. The study included evaluation of surface roughness and the process of chip formation. It was reported that a reduction in the specific cutting energy of 15.5% was achieved as a result of sharply increasing the cutting speed for very high values approaching 700%. A mere increment of one degree (1°) in the chip breaker bevel angle resulted in a 28.6% reduction in the specific cutting energy for normal speeds. The reduction becomes 13.7% for high-speed cutting (HSC). Values of workpiece roughness determined under various test conditions were very low so as to correspond to values resulting from conventional grinding operations, with a typical value around one fifth of a micrometer, or 0.2 μm.

In machining operations, the edge geometry of the cutting inserts has a great influence on process responses such as cutting forces, temperature, and surface roughness. There are two main geometries for cutting edges: conventional round-nose and wiper. The latter, wiper inserts, have been recently utilized to achieve a good surface quality in order to eliminate the need for further grinding process. Nevertheless, machining with wiper inserts could exhibit some negative impact in terms of higher cutting force and cutting temperature [7].

Mourão et al. [8] determined factors that significantly affect specific cutting energy (SCE) during face milling of aluminum alloys using wiper inserts. They reported that SCE is inversely proportional to the square of the cutter's cutting speed, indicating high sensitivity to changes in cutting speed. Additionally, SCE decreased as the depth of cut and feed per tooth increased, with the effect of the former more prominent than the latter.

Khan et al. [9] investigated the influence of microscopic geometry of wiper inserts on the resulting material hardness. They used two different geometries for the edge of wiper insert tools, the chamfer (0.15 mm × 25°) and the chamfer plus hone (0.15 mm × 25°-25 μm), in a hard-turning process of an AISI D2 steel alloy in dry conditions, i.e., without cutting fluid. They concluded that the role of wiper macro-geometry was somewhat suppressed in tool life as well as surface roughness. In particular, the hardness of the workpiece was the major factor affecting tool life (with a PCR value of 70.11%). With regard to surface roughness, the insert type (with wiper inserts showing significantly better effects than conventional tools) and feed rate played a larger role.

Dogra et al. [10] conducted a review of tool geometry variation entailing the radius of the tool nose, rake angle, rake face groove, variation of geometry of cutting edge geometry, geometry of wiper, and curved edge tools and their effects on tool wear/life, surface roughness, and also the integrity of machined surfaces. Moreover, the review included discussion of numerical approaches using modeling and simulation in tool geometry analysis, including an approach to variable micro-geometry tools developed in a recent study.

In another work, Abbas et al. [11] performed a comparative analysis between conventional and wiper inserts to investigate surface quality though a set of criteria (Ra, Rt, and Rz) in high-strength steel turning. The study highlighted the significance of the depth of cut and feed rate in improving surface roughness. An approach using the desirability function was employed to investigate the machining conditions leading to optimum surface roughness for the range of experimental results, in order to optimize multiple parameters of the response. The results revealed that the use of a wiper carbide insert produced significantly better surface quality than that produced using the conventional carbide insert. The improvement of the wiper insert over the conventional insert, which reached a maximum of 3.5-fold, was possible at a machining speed of 75 m/min. The improvements became lower as the speed was reduced, so were 3, 2.5, and 2 times at machining speeds of 100, 125, and 150 m/min, respectively.

In a following study, Abbas et al. [7] carried out an experimental evaluation of the surface quality produced during the precision turning process of the alloy steel AISI 4340 with conventional round and wiper inserts under various cutting conditions. An experimental design of full factorial with three parameters, each of them with four levels, was employed. The parameters are the feeding rate, the cutting speed, and the cutting depth. The resulting average roughness (Ra) is used to characterize the finished surface quality. The results showed that for the intended range of cutting conditions, wiper inserts produced lower surface roughness values, as opposed to conventional inserts, resulting in better surfaces. When the type of insert was included as a qualitative factor using ANOVA, it was found to be the most important factor for best surface roughness and metal removal rate. The feeding rate was the next factor of influence. Then, there came the interaction between feeding rate and insert type. Using wiper inserts made it possible to achieve a concurrent increase in feed rate, cutting depth, and cutting speed, and at the same time a superior quality of the resulting surface was obtained with a lower Ra value, as compared to surface roughness results when using normal cutting inserts. The improvements obtained through using wiper inserts over conventional ones reached up to ten times higher than metal removal rates. This is a clear indication of the enormous improvement in productivity achieved by wiper inserts over conventional inserts in precision hard turning of the alloy steel AISI 4340.

Processing AISI 1045 alloy steel samples by face milling was studied by Pimenov et al. to find the best processing conditions [12]. The study was inclusive of various parameters that included the cost of cutting tool components, consumption of energy, cutting tool wear, material removal rate, and, importantly, the resulting surface quality. Various experiments were conducted with variations in cutting length, after which the results were statistically studied in order to choose the optimum conditions of cutting. A multi-layer regression analysis was performed on the results of the experiments, which resulted in a nonlinear set of mathematical equations with a coefficient of determination of R2 = 0.98. The correlations considered in the study included the effects of the parameters feed per tooth (fz), speed of cut (vc), flank wear (VB) on surface roughness (Rz), material removal rate (MRR), sliding distance (ls), cutting performance (Pc), and last but not, least life of tool (T'). The overall results, estimated using Gray's relation analysis (GRA), showed that the optimum performance in fly milling for fast manufacturing (the first case) is obtained for feed per tooth fz = 0.25 mm/tooth, cutting speed vc = 392.6 m/min, and machined length l = 5 mm. In the second case, the optimal parameters for saving resources (mainly tools) were a feed per tooth of 0.125 mm/tooth, a cutting speed of 392.6 m/m, and a machined length of 5 mm.

Szczotkarz et al. [13] performed an assessment of the surface topography resulting from a turning process of the alloy steel AISI 1045 using carbide inserts with respect to the applied titanium-based coatings. The work presented the results of three-dimensional parameters, isometric views, contour maps, and material ratio curves. Analysis of the topography of the resulting surfaces revealed that for the TiAlN-coated insert at low cutting speeds and large feed rates, surface roughness parameters were low. In contrast, lower values of the selected 3D parameters resulted from the insert with TiC coating at higher cutting speeds. It was also reported that the TiC-coated insert produced the most uniform distribution of valleys and ridges in the machined surface. These results were used to determine the best ranges of cutting parameters, which allow appropriate selection of the type of titanium-based coating when machining this type of alloy.

D'Addona et al. [14] studied the surface roughness during hard turning with a wiper insert geometry. In the analysis, the surface roughness of wiper inserts and traditional inserts with a radius were compared. The analysis employed tools such as ANOVA, surface plots, and AOM. The conclusions of the study are as follows: wiper inserts produce better surface finish than traditional inserts with comparable surface finish during grinding. The feed rate proved to be the most important factor affecting the surface roughness. Additionally, feed rate then depth of cut and type of insert have statistically significant effects on surface roughness. The study concluded that the best processing conditions to produce high surface quality are as follows: a wiper nose radius of 1.2 mm, a speed of 1200 RPM, a feed rate of 0.08 mm/rev, and a depth of cut of 0.1 mm.

Patil et al. used the response surface method (RSM) to predict the process parameters in order to optimize the VMC-five axis milling of D3 steel. The responses selected for optimization in this study were the surface roughness and MRR. The multi-objective teaching learning-based optimization (MTLBO) technique was used to optimize the two combined responses. It was found that the machining conditions predicted by the hybrid RSM-MTLBO improved the studied responses significantly [15].

Jumare et al. used RSM to investigate the effects of three process parameters on the surface roughness (Ra) and tool wear in the diamond turning process of single-crystal silicon. ANOVA was used to examine the significance of the parameters (cutting speed, feed rate, and depth of cut) on the two responses. A desirability function multi-objective optimization approach was used to minimize both Ra and the tool wear, and to maximize MRR. The results showed that while the three parameters were significant, the feed rate had the largest effect on both responses [16].

Benkhelifa et al. built a mathematical model to optimize surface roughness and tool flank wear in the turning of AISI 316L stainless steel. The experiments were designed using the Taguchi L27 matrix. Both ANOVA and RSM techniques were applied to estimate the significance of the studied turning parameters, and to build a mathematical model for optimization. Multi-objective optimization using the desirability function was utilized in this study. The results showed that the feed rate was the main factor affecting both surface roughness and tool flank wear [17].

Rudrapati studied the individual and combined effects of grinding processing conditions on the surface quality of glass fiber reinforced epoxy composite. A full factorial design was used to plan the experimental work, and ANOVA was used to estimate the significance of each factor and interaction on the response. The desirability function was applied to predict the surface quality level as a function of quality level [18].

Looking at the reviewed literature, there has been a large number of investigations into the effect of different process parameters entailing the geometry of cutting inserts on individual or even a number of responses. Nevertheless, comprehensive studies of a wide range of parameters on several responses have not been fully covered. Therefore, the goal of this work is to perform an overall evaluation of the machinability performance of

AISI1045 when machined with a wiper in terms of the resulting surface, the cutting forces, and cutting temperature, and to compare these responses with those reported in the work of Abbas et al. [1] when conventional round-nose inserts were used.

2. Materials and Methods

As previously stated, this is a useful extension of the work reported by Abbas et al. in [1], in which the surface roughness, cutting force, and cutting temperature obtained when turning AISI1045 using conventional round-nose inserts were presented. AISI 1045 has played a key role as a reliable material for a number of applications such as gears, shafts, spindles, rollers, and crankshafts [1]. In this work, AISI 1045 samples are machined under uniform cutting conditions but with a different insert type with wiper geometry for the goal of assessing the machining performance of AISI1045 using different tool geometries. In addition, statistical analyses of the results for both cases are conducted and discussed.

2.1. Workpiece Materials

The material samples considered in the experiments are made from AISI 1045 steel, which is used in a wide range of applications in heavy industries where high strength and resistance to wear are desired. The specific elemental content of this alloy is shown in Table 1. The mechanical properties are provided in Table 2.

Table 1. Chemical content of AISI 1045 steel alloy [1].

Element	C	Mn	P	S	Fe
Percentage %	0.45	0.65	0.03	0.04	Balance

Table 2. Mechanical properties of AISI 1045 steel alloy [1].

Properties	Value
Ultimate Strength	565 MPa
Yield Stress	310 MPa
Fracture Elongation (in 50 mm)	16%
Area Reduction	40%
Elastic Modulus (Typical for steel)	200 GPa
Vickers Hardness	170

An optical microscope (OM) manufactured by Olympus, model: BX51-M, was used for the metallographic investigations. The samples were prepared according to the standard procedures for metallographic sample preparation. This includes grinding with SiC abrasive paper, subsequent polishing with a diamond paste of 1.0 and 0.05 μm, and final etching by immersion for 10 s in 5% Nital to visualize the microstructure of the sample. Figure 1a shows the optical micrograph, where the microstructure contains pearlite grains (light) in a ferrite matrix (dark). Pearlite grains consist of alternating lamellae of proeutectoid ferrite (Fe)/cementite (Fe3C) of random orientations, as shown in higher magnification in Figure 1b. The pearlite phase accounted for 43% of the volume, while the ferrite fraction was 57%.

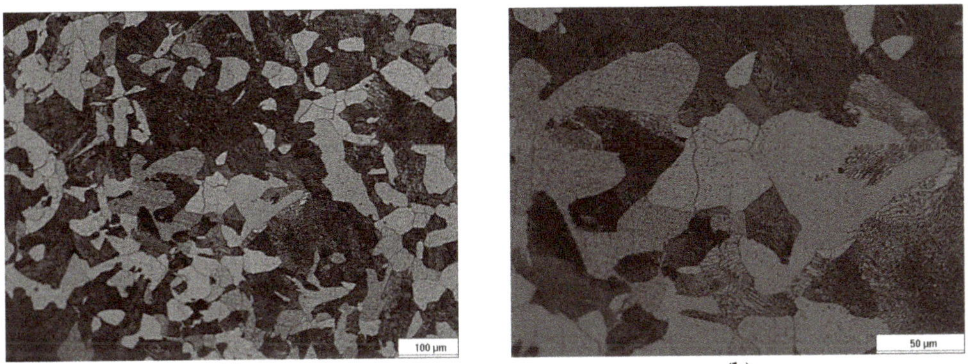

(**a**) (**b**)

Figure 1. Microstructure of AISI 1045 at two magnification levels, (**a**) 100 μm and (**b**) 50 μm.

2.2. Experimental Setup

A conventional lathe machine was used (type: EMCOMAT- 20D, from Emco Co., Salzburg, Austria) for machining test samples, see Figure 2. The machine specifications were 5.3 kw drive motor, with electronic speed control 40 to 3000 rpm, a longitudinal feed of 0.045 to 0.787 mm/rev, and stepless speeds.

Figure 2. Apparatus for machining the test samples and measuring the cutting forces and temperature.

The turning machine is a product of Sandvik (Stockholm, Sweden), with a holder of the type SDJCR 2020K 11 and an insert of the type DCMX11 T304- WF 4315 for the wiper cutting insert. This is compared to the conventional type insert used in the work reported in [1] with type number DCMT11 T304-PF 4315. The insert specifications were as follows: shape angle = 55°, clearance angle = 7°, rake angle = 6°, and tool nose radius = 0.4 mm. Optical microscope images of edge geometry for both conventional and wiper insert types are shown in Figure 3a,b. With design specifications for efficient material removal, this turning machine is utilized for various materials including stainless steel and aluminum and titanium alloys. The workpiece sample was 120 mm in length and 70 mm in diameter, with a cutting length of 30 mm for each round of experiments.

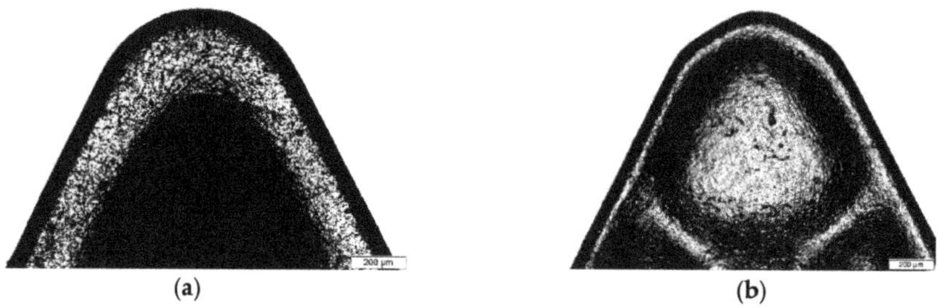

Figure 3. Optical microscope images of edge geometry: (**a**) conventional insert (**b**) wiper insert [7].

2.3. Design of Experiments

Full factorial design, with three factors, each of which had three levels, was used to build the experiment matrix. Specifically, the experimental plan was set with 27 test runs as follows: three levels of cutting speed 80, 120, and 160 m/min, three levels of cutting depth 0.5, 0.75, and 1.0 mm, and three levels of feed rate 0.045, 0.09, and 0.135 mm/revolution. Table 3 summarizes the factors and their levels. The full matrix is illustrated in the next section with the measured responses. Minitab 18 was used to check the significance of the three factors, and their interactions, on the responses through analysis of variance (ANOVA). ANOVA is a widely used technique to test the significance of factors and their interactions when more than two factors and/or interactions are examined. A 95% confidence level was set for the analysis, i.e., p-values below 0.05 prove the factor is significant and those above 0.05 show non-significance [19]. Backward elimination was applied to remove the non-significant items from ANOVA one at a time. Least squares multiple regression was utilized to build a model representing the three responses.

Table 3. Experimental cutting parameters with different levels for each parameter.

Designation	Process Parameter	Level 1	Level 2	Level 3
S	Cutting Speed (m/min)	80	120	160
D	Depth of Cut (mm)	0.50	0.75	1.00
FR	Feed Rate (mm/rev)	0.045	0.090	0.135

2.4. Characterization

For force measurement and calculation, a test stand of the type number Kistler 5070 was used with the software Dynoware 2825A (Liechtenstein, Switzerland) for data processing to calculate cutting force components: the main cutting force (F_c), the radial force (F_r), the feed force (F_f), and cutting force (F_c), see Figure 4. The resultant force (R) is evaluated using the standard relationship:

$$R = \sqrt{F_r^2 + F_f^2 + F_c^2} \tag{1}$$

The unit used in measurement of forces is Newton (N).

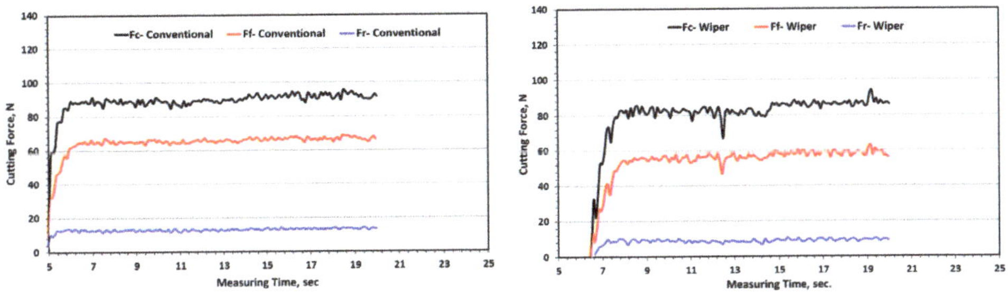

(a) (b)

Figure 4. Cutting forces Fr, Ff, and Fc for (**a**) conventional and (**b**) wiper inserts.

The material removal rate (MRR) is evaluated using the equation

$$\text{MRR} = 1000\, V\, f_r\, d \tag{2}$$

where V represents the surface speed (m/min.), f_r is the feed rate (mm/rev.), and d is the depth of cut (mm), and MRR is measured in (mm^3/min).

Thermal images were obtained with a ThermoPro-TP8 thermographic camera produced by Guide Co. (Wuhan, China). The camera must be stable and pointed at the target whose temperature is to be measured. In this case, the target is the contact between the cutting face of the insert tool and the surface of the sample during turning. The specifications of this camera are as follows: thermal sensitivity: ≤0.08 °C at 30 °C, measurement range: −20–1000 °C, detector type: micro-bolometer UFPA384 × 288 pixels, spectral range: 8∼14 µm, accuracy: ±2 °C. The parameters that must be specified include the distance of the target object from the camera lens and the emissivity, which must be specified according to the type of material, surface condition, temperature, and other factors. It is determined using a reference table in the camera manual. Care must be taken during calibration because although the camera performs automatic calibration, manual calibration must be performed before each exposure to achieve maximum sensitivity.

The test apparatus assembly for the machining process and measurements of cutting forces and temperature is shown in Figure 5.

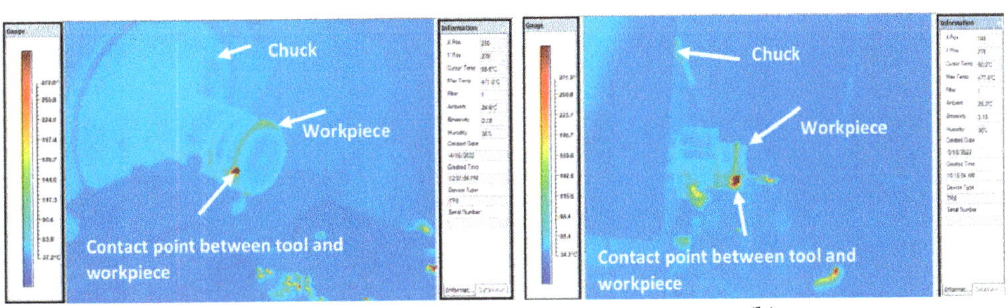

(a) (b)

Figure 5. Thermal image for (**a**) conventional and (**b**) wiper inserts.

For surface roughness (Ra) measurement, a Rugosurf 90-G type surface testing device from Tesa (Bugnon, Switzerland) was utilized. The measurement parameters were set as follows: cut-off length 0.8, cut-off number 15, and measurement speed 1 mm/s, and the measurements were performed on the curved surfaces.

3. Results and Discussion

3.1. Experimental Results

Figure 6 shows the obtainable surface roughness measurements for the swiper inserts versus the conventional inserts used under similar set conditions of cutting. The graph demonstrates the significant influence of the type of inserts used on the generated surface finish, which indicates the superiority of the wiper insert, resulting in a much lower surface roughness with a reduction percentage that varies between 31% and 60% in contrast to the results obtained from the case of conventional round-nose inserts under the same cutting parameters. To explain the superior performance of wiper inserts when compared with conventional round-nose ones, one should refer to the geometry of the wiper insert as presented in Figure 3. Specifically, while a conventional insert is composed of a single round nose, the wiper geometry consists of small radii that are well connected to the conventional nose. Therefore, the wiper edge geometry results in a nose with an effective straight section of minimal cutting nose. Therefore, with the aid of the straight section of the small cutting edge, when applying relatively small feed rates, the generated surface contains moderate peaks with a significantly reduced profile texture, resulting in improved surface quality [7].

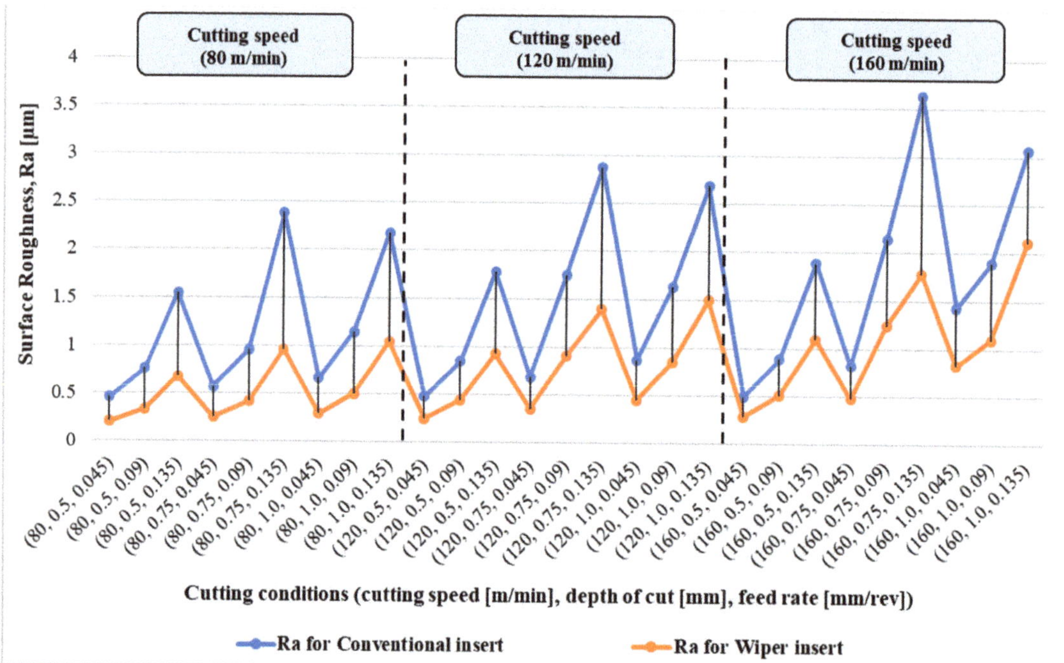

Figure 6. Resultant surface roughness, Ra in µm for both cases, of conventional and wiper inserts under a range of cutting conditions.

It is also not so difficult to see that the increase in the feed rate leads to a dramatic increase in the resulting surface roughness in both cases of tools, wiper and conventional round-nose inserts. In addition, looking at the surface roughness measurements achieved, one can say that higher cutting speeds have a noticeable negative effect on the final surface roughness for both insert types, wiper and conventional round-nose ones.

Figure 7 presents a comparison of the cutting forces obtained for the cases of wiper and conventional round-nose inserts for the full range of cutting conditions. The effects of cutting parameters on the resultant cutting forces can be summarized in two trends.

Firstly, the resultant cutting force is directly proportional to both feed rate and depth of cut. However, the cutting speed effect on the cutting force has a different trend. Specifically, for the wiper insert, the resultant cutting force is inversely proportional to the cutting speed, while for the conventional round-nose insert, the middle-range cutting speed of 120 m/min results in a higher cutting force when compared with other values of high and low cutting speeds. When looking at the effect of the insert type on the generated cutting force, a complex relationship can be concluded. In particular, the high and medium values of cutting speeds of the wiper inserts lead to the generation of less cutting force. However, at low cutting speeds, unclear trends are observed, where the resultant cutting forces for both cases are overlapping with minor deviations and flipping for different values of depth of cut. The results presented in Figure 7 show the nonlinear behavior of the turning process due to the effect of the cutting parameters or because of the type of cutting insert used.

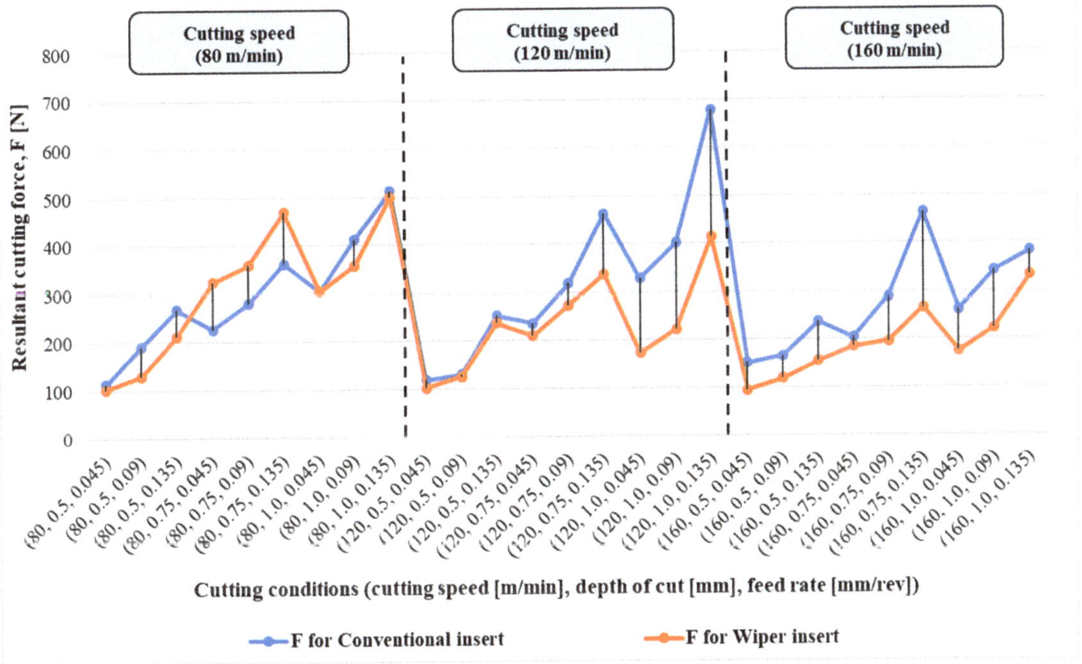

Figure 7. Resultant cutting force, F in N for both cases, of conventional and wiper inserts under a range of cutting conditions.

Figure 8 illustrates the cutting temperatures measured for both types of inserts under the entire range of cutting parameters. For the wiper insert, it is quite clear that cutting speed has a noticeable proportional effect on the detected thermal behavior of the process. Nevertheless, this is not the case for the conventional round-nose insert, where, except for the effect on feed and depth of cut at low cutting speed, there is no clear influence of the process parameters on the cutting temperature observed. The results also show the overlapping trend of the cutting temperature for both types of insert tools at low and medium values of cutting speeds, and a substantial difference is presented at high cutting speed, where the wiper inserts exhibit a higher thermal response of the process as opposed to the round-nose inserts.

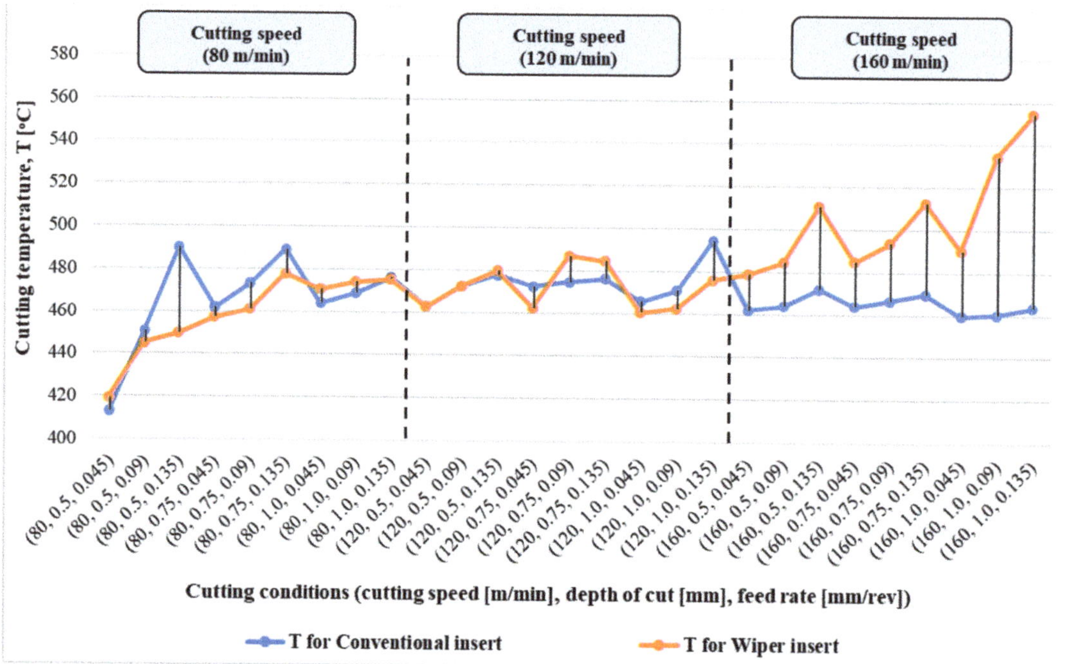

Figure 8. Cutting temperature, T in °C for both cases, of conventional and wiper inserts under a range of cutting conditions.

Looking at the results presented in Figures 6–8, it is not so difficult to see the contradicting effect of the type of insert, whether it is wiper or conventional round-nose inserts, on the three responses examined in this study. In particular, while surfaces produced with wiper inserts exhibit much less surface roughness compared to those from the conventional round-nose ones for the entire range of cutting conditions, the cutting forces resulting for the case of wiper inserts are higher than those associated with the round-nose ones, especially at high speeds of cutting. Furthermore, unclear and overlapping trends of the effect of the type of insert on the resultant cutting forces, particularly at low cutting speed, can be detected. However, even more crossed trends can be observed for the effect of the insert type on the obtainable cutting temperature at low and medium cutting speeds. Again, the aforementioned observation points out the need for an in-depth statistical analysis of the various effects of cutting parameters for each type of insert to elaborate further their effect and identify the optimal working window and cutting parameters for the best achievable process performance.

3.2. Statistical Analysis

3.2.1. Conventional Edge

Table 4 summarizes ANOVA results for Ra using the conventional round-nose cutting edge. The results show that the three investigated factors have significant effects on Ra. It is also shown that there is no significant interaction between these factors. Figure 9 illustrates the main effect plots of the three factors. The graph shows that the most significant effect on Ra comes from the feed rate, which is expected in turning processes.

Table 4. ANOVA analysis results for parameter Ra using conventional cutting edge.

Source	DF	Adj SS	Adj MS	F-Value	*p*-Value
S	1	1.738	1.7385	13.63	0.001
D	1	2.291	2.2905	17.96	0.000
FR	1	13.478	13.4784	105.68	0.000
Error	23	2.933	0.1275		
Total	26	20.441			

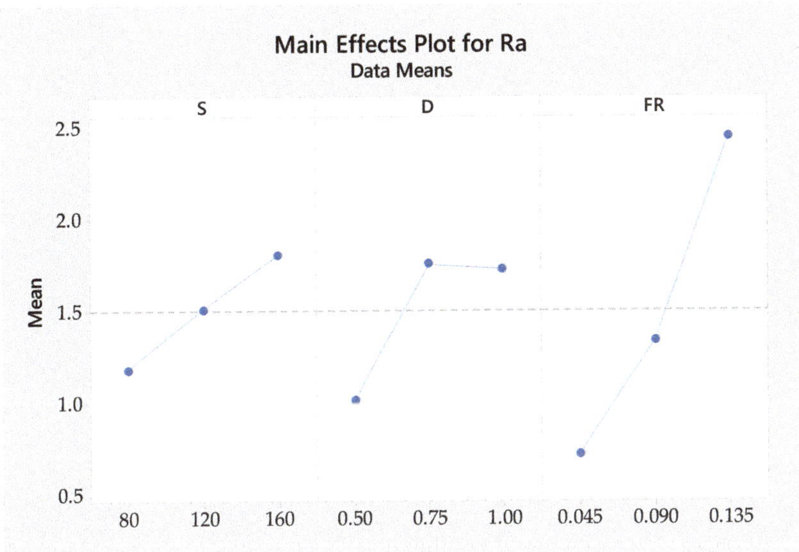

Figure 9. Main effect plots for Ra using conventional edge.

Table 5 shows ANOVA analysis for the resultant force using the conventional edge. The results show that only the feed rate and the cutting depth have a significant effect on the force. Again, this is expected in turning processes. The main effect plots of these two significant factors are depicted in Figure 10. The resultant force varies in almost the same pattern with both factors.

Table 5. ANOVA results for F using conventional cutting edge.

Source	DF	Adj SS	Adj MS	F-Value	*p*-Value
D	1	221,219	221,219	68.11	0.000
FR	1	156,572	156,572	48.21	0.000
Error	24	77,946	3248		
Total	26	455,737			

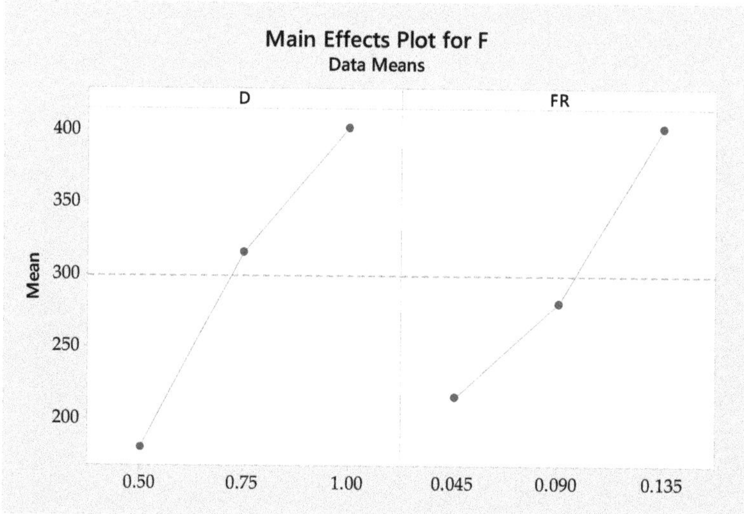

Figure 10. Main effect plots for F using conventional edge.

Table 6 presents the ANOVA results for the temperature of cutting using the conventional round-nose insert. The three studied factors affect the measured temperature significantly. Additionally, there is a significant interaction between the cutting speed and the cutting depth, and between the cutting speed and feed rate. Figure 11 shows the interaction and the contour plots for temperature vs. cutting speed and depth of cut. While the temperature increases continuously with increasing feed rate, it presents a different pattern to that of the cutting depth. The measured temperature increases for cutting depth values between 0.5 and 0.75 mm, after which it decreases slightly between 0.75 and 1.00 mm with cutting speeds of 80 and 120 m/min. Additionally, the rate of temperature increase with feed rate is steeper at the speed of 80 m/min than the other two speeds. The contour plot shows that the maximum temperature occurs at the maximum depth of cut and minimum cutting speed, while the minimum temperature occurs at minimum depth of cut and minimum cutting speed. It also presents a nonlinear distribution of temperature over the studied range of both factors.

Table 6. ANOVA results for T using conventional cutting edge.

Source	DF	Adj SS	Adj MS	F-Value	*p*-Value
S	1	981.8	981.8	8.64	0.008
D	1	538.8	538.8	4.74	0.041
FR	1	1467.8	1467.8	12.92	0.002
S*D	1	410.7	410.7	3.61	0.071
S*FR	1	779.2	779.2	6.86	0.016
Error	21	2386.3	113.6		
Total	26	5648.0			

The "*" sign represents interaction between the factors.

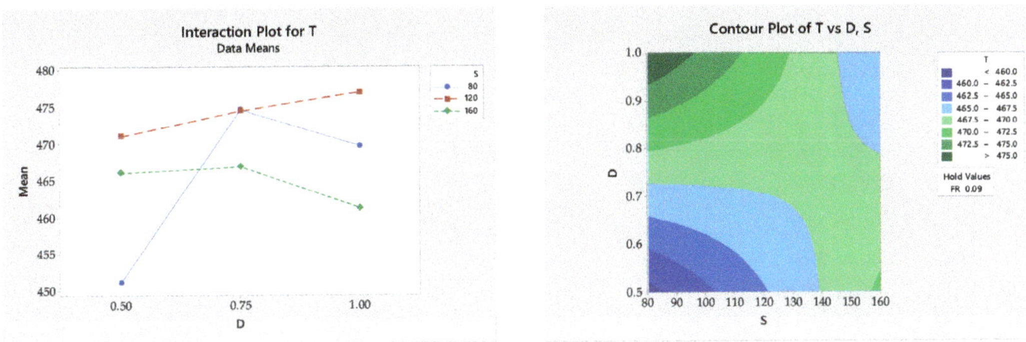

(a) (b)

Figure 11. (**a**) Interaction plot and (**b**) contour plot for T using conventional edge.

Table 7 summarizes the mathematical equations representing the measured outputs. These equations were constructed by regression using the least squares method. They could be used in predicting these responses under different machining conditions.

Table 7. Mathematical equations for the measured responses.

Response	Equation
Surface Roughness	Ra = −2.232 + 0.00778 S + 1.427 D + 19.23 FR
Cutting Force	F = −219.6 + 443.4 D + 2073 FR
Cutting Temperature	T = 337.5 + 0.836 S + 83.3 D + 764 FR − 0.585 S*D − 4.48 S*FR

The "*" sign represents interaction between the factors.

The desirability function optimization tool in Minitab 18 was used to minimize Ra, F, and T, and to maximize MRR. The priority was given to keeping Ra below 0.8 μm (with a target of 0.7 μm), then to maximize MRR and minimize T. The lowest priority was given to minimizing F. Individual desirability (d) and composite desirability (D) have a range of 0.0 to 1.0. One denotes the ideal case and zero illustrates the case when the response is outside its acceptable limits. The optimum values for the factors were calculated to be S = 160 m/min, D = 0.52 mm, and FR = 0.05 mm/rev. The expected optimum responses were calculated to be Ra = 0.7 μm, F = 114 N, T = 468 °C, and MRR = 3726 mm^3/min. The optimization plot is shown in Figure 12. For both force and surface roughness, d equals 1.0, proving that these two responses will be optimized with perfection. For temperature, d equals 0.84, and for MRR, d equals 0.7. These numbers show that the temperature and material removal rate will be optimized to a good extent, but not with perfection. For the combined responses, D, equals 0.87, proving that the combined optimization is well achieved.

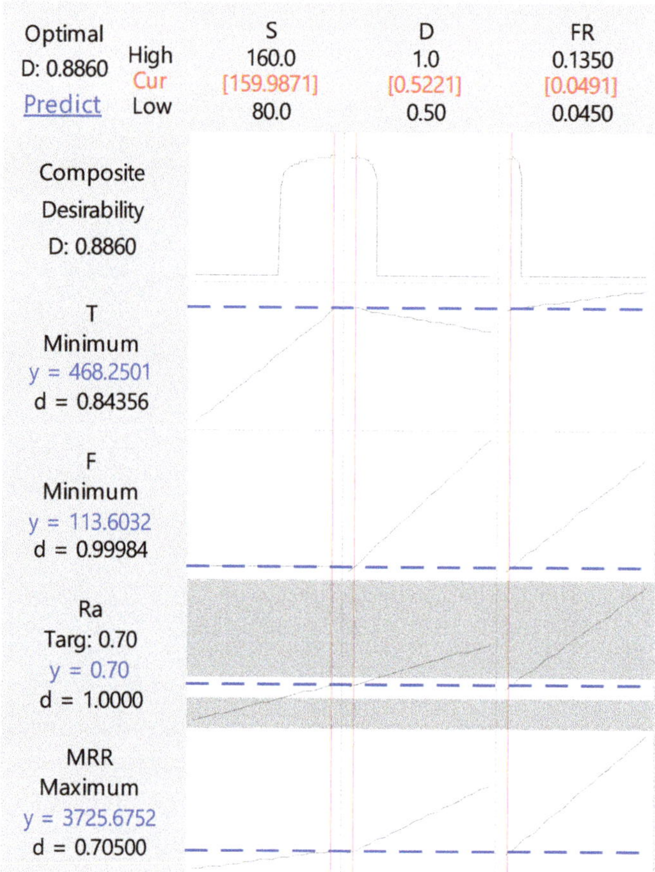

Figure 12. Optimization plot for responses using conventional cutting edge.

3.2.2. Wiper Edge

Table 8 summarizes the ANOVA results for Ra using the wiper cutting edge. The results show that the investigated three factors have significant effects on Ra, in a similar pattern to that of the conventional edge. Figure 13 illustrates the main effect plots of the three factors. It is clear that the wiper cutting edge results in higher surface roughness, with almost half of the conventional cutting edge values.

Table 8. ANOVA results for Ra using wiper cutting edge.

Source	DF	Adj SS	Adj MS	F-Value	*p*-Value
S	1	1.2461	1.24609	34.80	0.000
D	1	0.8642	0.86417	24.14	0.000
FR	1	3.6450	3.64500	101.80	0.000
Error	23	0.8235	0.03580		
Total	26	6.5788			

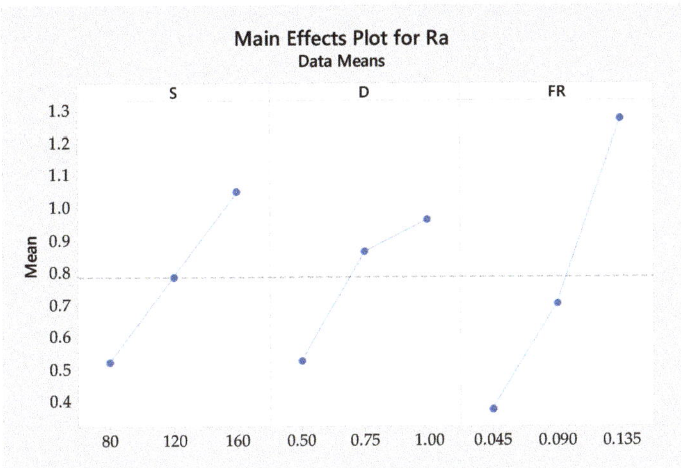

Figure 13. Main effect plots for Ra using wiper cutting edge.

The results from ANOVA analysis for resultant force using the wiper cutting edge tool are displayed in Table 9. It is clear from the table that all three factors have significant effects on the resultant force. These results are different from the conventional cutting edge, in which the cutting speed effect on the resultant force was not significant. The main effect plots of the three factors are shown in Figure 14.

Table 9. ANOVA results for F using wiper cutting edge.

Source	DF	Adj SS	Adj MS	F-Value	*p*-Value
S	1	57,529	57,529	17.77	0.000
D	1	112,689	112,689	34.81	0.000
FR	1	87,185	87,185	26.94	0.000
Error	23	74,447	3237		
Total	26	331,850			

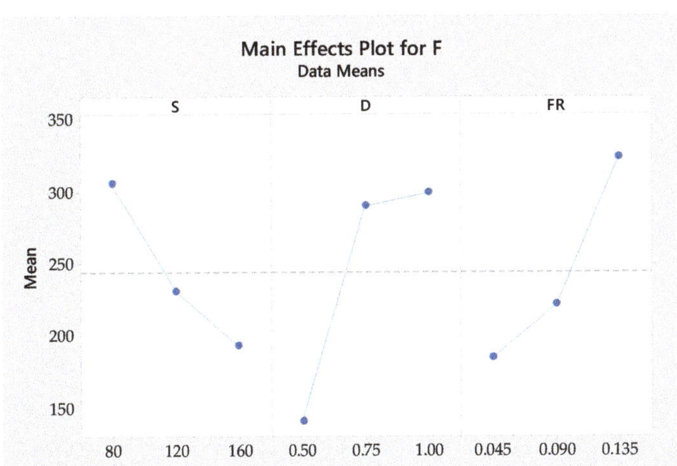

Figure 14. Main effect plots for F using wiper edge.

Table 10 summarizes the ANOVA results for the measured temperature with the wiper cutting edge. The three factors have significant effects with no interaction, in contrast to the conventional edge. Figure 15 shows the main effect plots for the temperature.

Table 10. ANOVA results for T using wiper cutting edge.

Source	DF	Adj SS	Adj MS	F-Value	*p*-Value
S	1	9587	9587.4	50.38	0.000
D	1	2123	2123.3	11.16	0.003
FR	1	3044	3044.1	15.99	0.001
Error	23	4377	190.3		
Total	26	19,132			

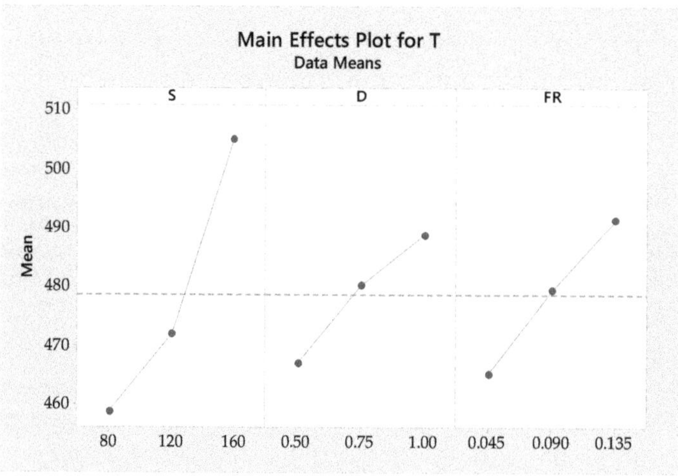

Figure 15. Main effect plots for T using wiper cutting edge.

Table 11 summarizes the mathematical equations representing the measured outputs. These equations were constructed by regression using the least squares method. They could be used in predicting these responses under different machining conditions.

Table 11. Mathematical equations for the measure responses.

Response	Equation
Surface Roughness	Ra = −1.56 + 0.00658 S + 0.876 D + 10.00 FR
Cutting Force	F = −37.0 + 316.5 D + 1547 FR
Cutting Temperature	T = 350.9 + 0.577 S + 43.4 D + 289 FR

The desirability function optimization tool was used to minimize Ra, F, and T, and to maximize MRR with the same optimizing conditions used with the conventional edge. The optimum values for the factors were calculated to be S = 160 m/min, D = 0.52 mm, and FR = 0.075 mm/rev. The expected optimum responses were calculated to be Ra = 0.7 µm, F = 93.7 N, T = 487.5 °C, and MRR = 6163 mm^3/min. Figure 16 shows the optimization plot.

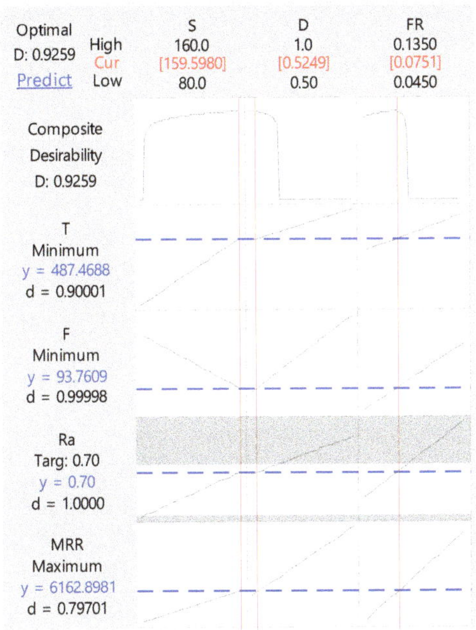

Figure 16. Optimization plot for responses using wiper cutting edge.

For both force and surface roughness, d equals 1.0, proving that these two responses will be optimized with perfection. For temperature, d equals 0.9, and for MRR, d equals 0.79. These numbers show that the temperature and material removal rate will be optimized to a good extent, but not with perfection. For the combined responses, D, equals 0.93, proving that the combined optimization is well achieved. The results show that the same value of Ra could be achieved with the wiper edge at a higher removal rate than the conventional edge, which gives an advantage for the wiper edge in machining.

4. Conclusions

This manuscript presented an experimental investigation and statistical analysis of the dry turning of the AISI 1045 steel alloy for two cases of tools inserts used in machining: the wiper type insert and the conventional round-nose insert. The experimental results for the performance of the wiper inserts are compared with the results recently reported by the authors in [1] and obtained under similar cutting conditions. In particular, the following is concluded:

- For conventional inserts, the optimal process conditions to be applied are cutting speed: 160 m/min, cut depth: 0.52 mm, and feed rate: 0.05 mm/rev. These parameters produce a surface roughness of 0.7 µm, a cutting force of 114 N, a cutting temperature of 468 °C, and a material removal rate of 3726 mm^3/min.
- For a wiper-shaped insert, the optimal parameters are a cutting speed of 160 m/min, depth of cut 0.52 mm, and feed rate 0.075 mm/rev, which produce the following optimum responses: surface roughness: 0.7 µm, cutting force: 93.7 N, cutting temperature: 487.5 °C, and material removal rate: 6163 mm^3/min.
- One can conclude that cutting inserts with a wiper edge prove to provide lower surface roughness than inserts with a conventional cutting edge under the same set of cutting conditions. This advantage allows wiper tools to use higher feed rates, resulting in a greater material removal rate, while obtaining the same value of surface roughness, which gives an advantage for the wiper edge in machining.

- There are no notable differences between the resultant cutting forces of both cutting edges at low cutting speeds. However, for higher cutting speeds, wiper inserts outperform the conventional round inserts with regard to obtainable cutting forces, which demonstrates the superiority of wiper inserts not only in terms of surface roughness but for less cutting force.
- At high cutting speeds, wiper inserts exhibit higher values of cutting temperature and, thus, conventional round-nose inserts are recommended at such high levels of cutting speeds.
- For conventional round-nose inserts, the feed rate was found to be the most significant parameter affecting the generated surface roughness, cutting force, and temperature. For wiper inserts, the feed rate was found to be the most significant parameter affecting surface roughness and cutting speed, while for cutting temperature, cutting speed was the most effective factor.

Author Contributions: Concept, A.T.A. and A.E; methodology, A.T.A. and A.E.; software, A.A.A.-A. and F.B.; validation, A.A.A.-A., M.M.E.R. and A.E.R.; formal analysis, A.E., A.E.R. and A.T.A.; investigation, A.E., A.E.R., A.A.A.-A. and F.B.; resources, A.T.A.; data curation, A.T.A., A.E.R., M.M.E.R. and A.A.A.-A.; original draft preparation, A.E.R., A.T.A. and A.E.; writing—review and editing, A.A.A.-A., F.B. and A.E.; visualization, F.B., A.E. and A.A.A.-A.; supervision, A.E. and A.T.A.; project administration, A.A.A.-A. and F.B.; funding acquisition, A.T.A. All authors have read and agreed to the published version of the manuscript.

Funding: This research received no external funding.

Data Availability Statement: Not applicable.

Acknowledgments: The authors acknowledge the funding of the Deanship of Scientific Research at King Saud University through research group No. RG-1439-020 with appreciation.

Conflicts of Interest: The authors declare no conflict of interest.

References

1. Abbas, A.T.; Al-Abduljabbar, A.A.; El Rayes, M.M.; Benyahia, F.; Abdelgaliel, I.H.; Elkaseer, A. Multi-Objective Optimization of Performance Indicators in Turning of AISI 1045 under Dry Cutting Conditions. *Metals* **2023**, *13*, 96. [CrossRef]
2. Abbas, A.T.; Al-Abduljabbar, A.A.; Alnaser, I.A.; Aly, M.F.; Abdelgaliel, I.H.; Elkaseer, A. A Closer Look at Precision Hard Turning of AISI4340: Multi-Objective Optimization for Simultaneous Low Surface Roughness and High Productivity. *Materials* **2022**, *15*, 2106. [CrossRef] [PubMed]
3. Abbas, A.T.; Gupta, M.K.; Soliman, M.S.; Mia, M.; Hegab, H.; Luqman, M.; Pimenov, D.Y. Sustainability assessment associated with surface roughness and power consumption characteristics in nanofluid MQL-assisted turning of AISI 1045 steel. *Int. J. Adv. Manuf. Technol.* **2019**, *105*, 1311–1327. [CrossRef]
4. Brown, I.; Schoop, J. The effect of cutting edge geometry, nose radius and feed on surface integrity in finish turning of Ti-6Al4V. *Procedia CIRP* **2020**, *87*, 142–147. [CrossRef]
5. Khidhir, B.A.; Mohamed, B. Analyzing the effect of cutting parameters on surface roughness and tool wear when machining nickel based hastelloy—276. *Mater. Sci. Eng.* **2011**, *17*, 012043. [CrossRef]
6. Rodrigues, A.R.; Coelho, R.T. Influence of the Tool Edge Geometry on Specific Cutting Energy at High Speed Cutting. *J. Braz. Soc. Mech. Sci. Eng.* **2007**, *29*, 279–283. [CrossRef]
7. Abbas, A.T.; El Rayes, M.M.; Luqman, M.; Naeim, N.; Hegab, H.; Elkaseer, A. On the Assessment of Surface Quality and Productivity Aspects in Precision Hard Turning of AISI 4340 Steel Alloy: Relative Performance of Wiper vs. Conventional Inserts. *Materials* **2020**, *13*, 2036. [CrossRef] [PubMed]
8. Mourão, A.; Slătineanu, L.; Gonçalves-Coelho, A.M. The effect of wiper edge inserts on the specific cutting energy in face milling of aluminium alloys. *Int. J. Mod. Manuf. Technol.* **2010**, *2*, 71–77.
9. Khan, S.A.; Umar, M.; Saleem, M.Q.; Mufti, N.A.; Raza, S.F. Experimental investigations on wiper inserts' edge preparation, workpiece hardness and operating parameters in hard turning of AISI D2 steel. *J. Manuf. Process.* **2018**, *34*, 187–196. [CrossRef]
10. Dogra, M.; Sharma, V.S.; Dureja, J. Effect of tool geometry variation on finish turning—A Review. *J. Eng. Sci. Technol. Rev.* **2011**, *4*, 1–13. [CrossRef]
11. Abbas, A.T.M. Comparative Assessment of Wiper and Conventional Carbide Inserts on Surface Roughness in the Turning of High Strength Steel. *J. Mater. Sci. Res.* **2016**, *5*, 32.

12. Pimenov, D.Y.; Abbas, A.T.; Gupta, M.K.; Erdakov, I.N.; Soliman, M.S.; El Rayes, M.M. Investigations of surface quality and energy consumption associated with costs and material removal rate during face milling of AISI1045 steel. *Int. J. Adv. Manuf. Technol.* **2020**, *107*, 3511–3525. [CrossRef]
13. Szczotkarz, N.; Maruda, R.W.; Debowski, D.; Leksycki, K.; Wojciechowski, S.; Khanna, N.; Królczyk, G.M. Formation of Surface Topography During Turning of AISI 1045 Steel Considering the Type of Cutting Edge Coating. *Adv. Sci. Technol. Res. J.* **2021**, *15*, 253–266. [CrossRef]
14. D'Addona, D.M.; Raykar, S.J. Analysis of surface roughness in hard turning using wiper insert geometry. *Procedia CIRP* **2016**, *41*, 841–846. [CrossRef]
15. Patil, A.; Rudrapati, R.; Poonawala, N.S. Examination and prediction of process parameters for surface roughness and MRR in VMC-five axis machining of D3 steel by using RSM and MTLBO. *Mater. Today Proc.* **2021**, *44*, 2748–2753. [CrossRef]
16. Jumare, A.I.; Abou-El-Hossein, K.; Abdulkadir, L.N.; Liman, M.M. Predictive modeling and multiobjective optimization of diamond turning process of single-crystal silicon using RSM and desirability function approach. *Int. J. Adv. Manuf. Technol.* **2019**, *103*, 4205–4220. [CrossRef]
17. Benkhelifa, O.; Cherfia, A.; Nouioua, M. Modeling and multi-response optimization of cutting parameters in turning of AISI 316L using RSM and desirability function approach. *Int. J. Adv. Manuf. Technol.* **2022**, *122*, 1987–2002. [CrossRef]
18. Ramesh, R. Prediction of surface roughness in cylindrical grinding of glass fiber reinforced epoxy composite. *Int. J. Mach. Mach. Mater.* **2022**, *24*, 405–418.
19. Phanphet, S.; Bangphan, S. Application of Full Factorial Design for Optimization of Production Process by Turning Machine. *J. Tianjin Univ. Sci. Technol.* **2021**, *54*, 35–55.

machines

MDPI

Review

Elucidating Powder-Mixed Electric Discharge Machining Process, Applicability, Trends and Futuristic Perspectives

Iqtidar Ahmed Gul [1,*], Ahmad Majdi Abdul-Rani [1], Md Al-Amin [2] and Elhuseini Garba [1]

1 Department of Mechanical Engineering, Universiti Teknologi PETRONAS, Seri Iskandar 32610, Malaysia
2 School of Mechanical and Mining Engineering, The University of Queensland, St Lucia, QLD 4072, Australia
* Correspondence: iqtidar_21002720@utp.edu.my

Abstract: Since the inception of electric discharge machining (EDM), it has facilitated the production industries, for instance, die & mold, automotive, aerospace, etc., by providing an effective solution for machining hard-to-cut materials and intricate geometries. However, achieving high machining rates and a fine surface finish is an inherent issue with the traditional EDM process. The emergence of the powder mixed electric discharge machining (PMEDM) process has not only provided the opportunity for enhancing productivity and surface finish but also opened a window for its potential application in surface modification/coating of biomaterials. The process incorporates simultaneous machining and coating of bioimplants, i.e., lacking in the already available chemical and physical coating methods while requiring costly post-treatment procedures. This study comprehends the influence of powder characteristics and EDM process parameters on the performance parameters. The impact of tool electrodes and additive powders on the machined and coated surface of commonly used biomaterials. Furthermore, the study depicts the most frequently used methods for optimizing the PMEDM process, future research directions, challenges, and research trends over the past decade.

Keywords: EDM; PMEDM; surface modification; powders; coating

1. Introduction

Electric Discharge Machining (EDM) technology has been applied efficiently in the die and mold-making industries, focusing on component precision rather than increased production [1–3]. EDM is an electrothermal material erosion process that primarily uses electrical energy and converts it into thermal energy through a cycle of discontinuous electrical discharges ongoing among the tool and workpiece submerged in a dielectric medium. The thermal energy produces a plasma channel between the anode and cathode at extremely high temperatures (8000 °C to 12,000 °C) and pressure (20 MPa). During the machine work, the materials melt and evaporate in the presence of a dielectric which fulfils the requirement of coolant, insulation and flushing of microscopic debris [4,5]. EDM can efficiently machine hard, ductile, and brittle materials and create geometrically complicated shapes since there is no direct connection between the electrode and the workpiece [6]. Nevertheless, the significant constraints, for instance, low material removal rate, inadequate surface quality, time-taking, and restriction to machine-only conductive materials, mitigate the employment of EDM [7–10].

The innovation of powder-mixed electric discharge machining (PMEDM) occurred in the late 1970s, and the initial publication was attained in the 1980s [11]. Mitsubishi originated the revolutionary dual tanks for EDM in the additive process marketplace [12]. The tank containing a dielectric only initiates the rough machining process, while the other tank comprising a powder-mixed dielectric provides the finish machining process. PMEDM is an innovative surface modification technique that can simultaneously machine and coat the workpiece and addresses the limitations to significant extent that exist in EDM. In PMEDM, thermally and electrically conductive powders are mixed with a dielectric fluid to strengthen the machining performance measures and altered surface characteristics by lowering the shielding strength of the dielectric fluid and augmenting the sparks among the tool and the workpiece. The gap between the electrode and workpiece was reported to be increased by 200%, enlarging the plasma channel and producing a shallow crater on the machined surface [13]. This technological advancement draws researchers' attention to significantly considering machined surfaces with alloy modification. A standard PMEDM arrangement, as shown in Figure 1, is a slightly re-designed and upgraded version of the EDM setup.

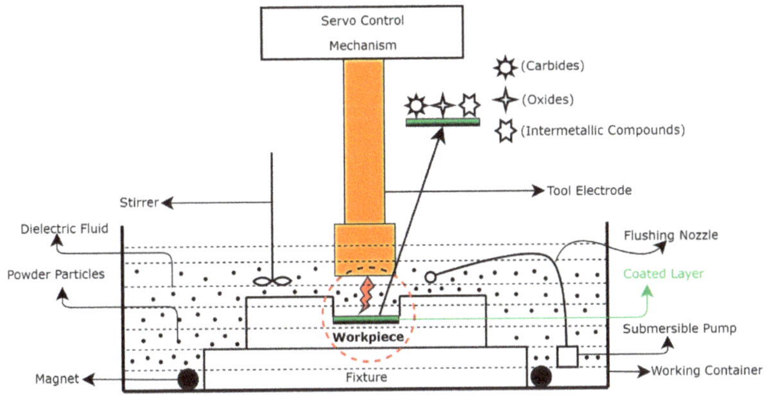

Figure 1. Typical arrangement of PMEDM system.

In the PMEDM process, powder particles are suspended in a dielectric fluid with a powder circulation system via a stirrer and a set of permanent magnets to filter the remnants. Usually, a submersible pump is equipped to prevent the settlement of powder bits at the base of the working container. It also assists in avoiding the resting of powder particles on the component surface. A pair of permanent magnets enable the detachment of residues from the powder particles, which only applies when the workpiece is magnetic, and the powder material is non-magnetic. The add-on of appropriate powder particles to the dielectric contributes to a superior surface texture and improved production capability compared to the EDM that excludes powder-mixed dielectrics. The industrialization of PMEDM still requires a breakthrough. It can be accomplished by having a complete insight into its process mechanism, correlation, and influence of process factors on the performance responses.

The prime objective of this literature review study is to elucidate the PMEDM mechanism, the impact of powder characteristics and machining parameters on the PMEDM outcome and its potentiality in biomedical applications. Secondly, exploring the trend of additives employed in the PMEDM process, future research directions in the published literature and probing the opportunities for its commercialization.

2. Mechanism of PMEDM Process

The mechanism of PMEDM is yet to be explored, which involves the suspension of fine powder particles in the dielectric medium. An electrical field of 10^5–10^7 V/m is produced in the inter-electrode gap (IEG) of around 25–50 µm when adequate voltage (nearly 80–320 V) is used among the tool electrode and substrate. The powder particles get activated and build up in a random pattern, as illustrated in Figure 2b. Under the sparking region, particles form chain-like formations between the electrode and substrate. The interlinking of the powder particles occurs in the current flow path. The chain-like arrangement of the molecules assists in bridging the spark gap between the electrode and the substrate and eventually reduces the insulating strength of the dielectric. The short-circuiting phenomenon causes a premature explosion in the IEG area, resulting in a series of discharges with rapid sparking and enhanced material removal rate [14]. Concurrently, the added powder particles augment and broaden the discharge gap and passage with uniformly distributed sparking between the particles. It reduces spark current density, resulting in constant erosion with a finer finish on the machined surface [15].

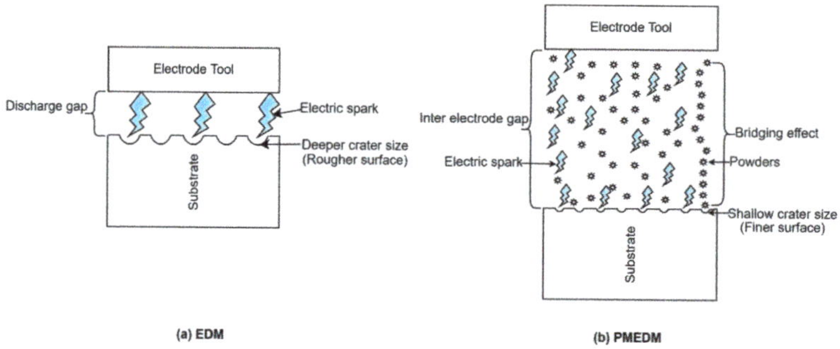

Figure 2. (**a**) Sparking mechanism in electric discharge machining process; (**b**) Sparking mechanism in powder-mixed electric discharge machining process.

Mixing powder particles in the dielectric medium enables the enlargement and expansion of the machining gap, even the dispersal of sparks and alloying surface layer. The effect of Al powder particles in the kerosene dielectric medium during the machining operation is displayed in Figure 3. From the representation, it can be observed that Figure 3a depicts the sparking procedure of conventional EDM. In contrast, the steps in Figure 3b–g represent the PMEDM process with an enlarged and widened discharge gap resulting in the formation of tiny craters due to the multiple discharges and bridging arrangement between the added Al particles. Figure 3h shows the voltage waveform using the kerosene with Al powder particles during the single pulse duration [13].

The electrical and physical characteristics of the powder particles portray a significant impact on the increase of the discharge gap. The available electrons lying in electrically conductive powders lessen the insulating strength of the dielectric. The enhanced conductivity results in the development of a spark from an enlarged distance and thus augments the discharge gap [16]. Following the initial discharge, powder particles in IEG become excited and proceed promptly in conjunction with ions and electrons. These energetic powder particles impact dielectric molecules and trigger further ions and electrons [17]. Therefore, a higher number of electrical charges are generated in PMEDM compared to traditional EDM.

Additionally, hydrostatic pressure operating on the plasma channel is lessened due to the discharge gap augmentation. The occurrences of these two incidents endorse the broadening of the discharge path. The severity of discharge strength lowered due to the increased and broadened discharge column, leading to substantial shallow cavities on the machined surface.

In a single pulse-on duration, a series of discharge pathways are examined in PMEDM caused by the abrupt zigzag displacement of the added powder particles, securing consistent energy dispersion and creating several minor craters. Unlike EDM, several fluctuating signals report various discharges in the PMEDM process during an individual pulse-on time [18,19]. When the dielectric fluid disintegrates, the carbon and oxygen molecules separate from it and blend with the partly melted powder particles and the residues of molten metal to shape carbides and oxides. Due to the exceptional amount of powder particles, the enervated debris could not flush out properly and suppresses in between the discharge gaps resulting in the relocation towards the substrate surface and infiltrating the molten pool until the curing occurs [20]. Succeeded by additional discharges, the carbides and oxides speed up, generate negative pressure by electrophoresis, and migrate towards the machined surface. The coupled layer of carbides and oxides in the shape of micro-droplets develops a consolidated region on the upper surface of the machined substrate. Favorably, in parallel, an immense quantity of absorbed gas is expelled, which results in the making of pores and a smoother shape on the upper surface of the machined component, as stated by Peng et al. [21].

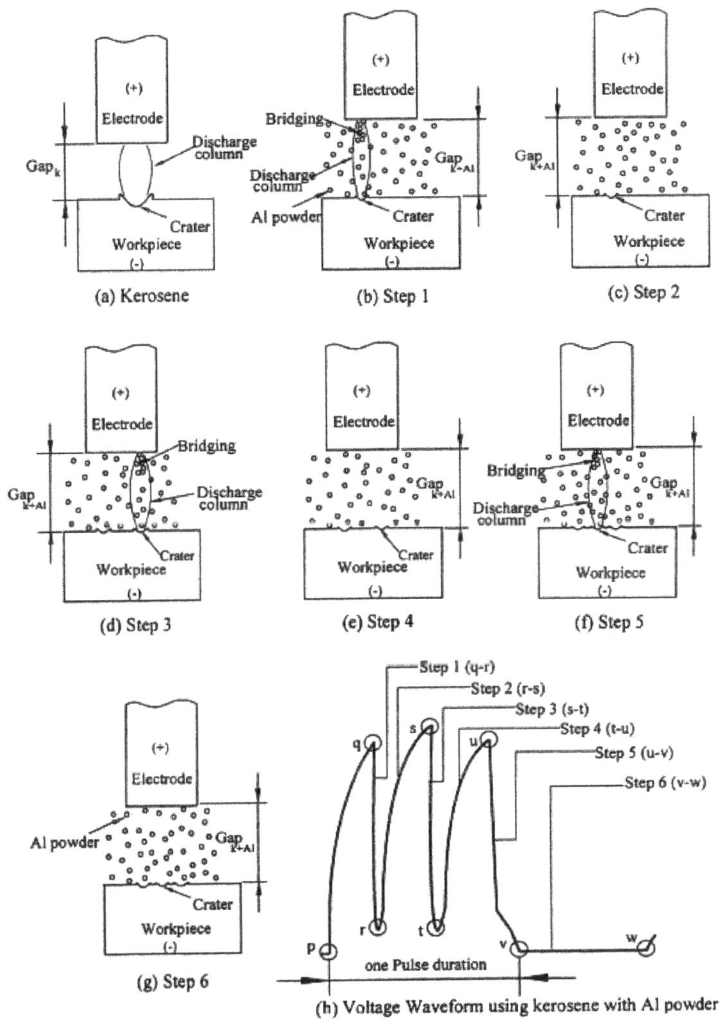

Figure 3. Influence of Al powder additive in kerosene dielectric medium during the machining process [22].

The migrated coupled layer of oxides and carbides decreases micro-cracks generation on the modified coated surface and enhances the mechanical characteristics [23]. The fundamental structure of material migration throughout the PMEDM process is demonstrated in Figure 4. The schematic description of material migration during the PMEDM process is demonstrated in the subsequent steps: Step 1 indicates that the sparking develops the plasma channel of thermal flux among the powder particles and substrate surface; Step 2 demonstrates the assemblage of hydrocarbon (C) and oxygen bubbles (O) among the partly dissolved powder particles and molten metal micro-droplets (M); Step 3 shows the arrangement of carbides (MC, XC) and oxides (MO2, XO2) due to the existence of hydrocarbon (C), oxygen molecule (O), and powder particles (X) and Step 4 exhibits the infiltration of oxides and carbides in the molten pool area of the coated layer before curing on the substrate surface in PMEDM method. Janmanee et al. [23] examined the deposition of a tough layer of tungsten carbide by EDM technique using titanium as a powder additive. The migrated, coupled layer diminishes the micro-cracks on the machined surface of the workpiece and enhances the mechanical features.

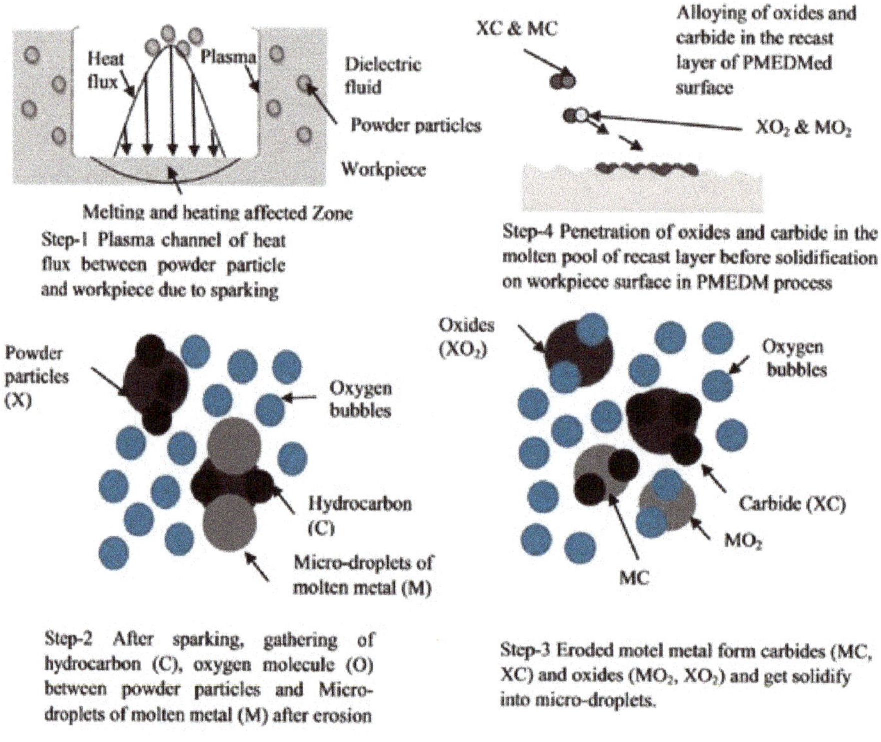

Figure 4. Schematic demonstration of material migration during the PMEDM process [24].

High heat generation and rapid quenching during the PMEDM process result in electrode materials melting, evaporation and re-solidifying. Due to this, surface modification naturally occurs in the PMEDM process with the stochastic behavioral response of process parameters.

3. Powder Characteristics and Its Utilization in the PMEDM Process

In the PMEDM process, the powder features play a crucial role in impacting several performance actions of the process. Therefore, it is imperative to assess powder characteristics and their interconnectivity with the EDM performance measures. Powder material, size and concentration are mainly essential features of the powder particles that influence the PMEDM process.

3.1. Powder Material

The essential thermophysical properties of the powder, i.e., particle density and electrical and thermal conductivity, significantly impact the performance measures of the PMEDM process. Wong et al. [25] investigated the outcome of Graphite (Gr), Silicon (Si), Aluminum (Al), Silicon carbide (SiC), Molybdenum disulphide (MoS2) and crushed glass on PMEDM performance measures. They discovered that, except for crushed glass, all powders increased the discharge gap by at least two times and provided the near-mirror surface finish in machining SKH-51. The superior electrical and thermal conductivity of Al powder provided the major machining gap expansion by nearly 12 times and the best surface quality. The gap expansion attributes for Al powder are in harmony with the research by Chow et al. [22] that assures the widening of the discharge gap and superior surface finish by adding Al powder particles to the dielectric medium. However, the influence of SiC additive is not in agreement because of its high electrical resistivity. Tzeng et al. [26] pointed out that within Al, SiC and Cr powder particles, Cr provided the maximum material removal rate, followed by Al and SiC. In comparison, the tool wear rate was noticed with SiC, Al and Cr, respectively. Yih-Fong and Fu-Chen [27] examined the Al, Cr, SiC and Cu powder particles and indicated that Al powder established the best surface finish and thinner white layers on the machined surface of SKD-11. Copper does not participate in the process due to its high particle density, which causes the powder to settle into the tank's bottom. Adding Graphite (Gr) powder particles to the dielectric increases its electrical conductivity and offers exceptional lubricity. The suspended Gr powder particles deliver enhanced material erosion rate and lowered tool erosion rate, and its auspicious application lies in the micro-PMEDM [16,28,29]. The utilization of Titanium (Ti) as an additive powder in the dielectric enhances the material erosion rate [30], micro-hardness [23], and surface quality. In contrast, it lowers the surface roughness and micro-cracks on the machined product. Likewise, Tungsten (W) powder-mixed dielectric boosts the micro-hardness of the machined surface by 100% [31].

3.2. Powder Size

The size of the powder particle is considered a vital powder parameter to get desirable outcomes during the PMEDM process. Utilizing smaller powder particles in the dielectric fluid improves the metal erosion rate and surface condition of the modified surface compared to the larger powder particles [26,32–34]. According to the research by Tzeng and Lee [26], smaller particles (70–80 nm) develop a minimal machining gap between the tool and the substrate, resulting in a more significant material removal rate and lower tool wear rate. Research by Yih-Fong et al. [27] reported that the size or dimension of the powder particle is the decisive factor in the machined surface quality. Smaller particles result in a superior surface finish while augmenting the white layer thickness.

3.3. Powder Concentration

A suitable quantity of powder drives effective machining operations and stability. A higher amount of powder is useful in adding the number of discharges that increases the material removal rate [35]. Increasing the powder concentration over the optimum value results in arching, short-circuiting, and unsteady machining due to unnecessary powder and residue particles [26]. Moreover, the existence of excessive powder particles in the machining gap results in surface degradation and settlement issues [36]. The estimation of powder concentration is crucial as its low value will not significantly influence the process dynamics and hence the outcome responses. Contrarily, the higher quantity of powder can induce particle settling, unrestrained bridging effects, and unsteady machining (short-circuiting and arcing), resulting in the degradation of the altered surfaces.

3.4. Powder Additives Utilization and Trend

Powder additives are added to the PMEDM process mainly for (i) to enhance the process efficiency and (ii) to upgrade the EDM process to surface coating technology. In the case of process efficiency, the prime process characteristics are material removal rate (MMR), tool wear rate (TWR), and inter-electrode gap (IEG). Subsequently, the surface modification process can improve microhardness, adhesion strength, coated layer thickness, surface roughness (SR), biocompatibility, wettability, corrosion and wear resistance. Table 1 lists the powder additives used in the PMEDM process over the last decade and their purpose of utilization.

Table 1. Powders utilized in the PMEDM process over the last decade.

Powders	Nomenclature	Impact on the Process Efficiency and Effectiveness
Graphite	Gr	Increases electrical conductivity that results in high MRR and low TWR
Aluminium	Al	High surface quality, enhances fatigue endurance and corrosion resistance
Silicon	Si	Develops a mirror-like surface finish on the machined surface
Silicon carbide	SiC	High MRR and surface integrity with the formation of hard layers
Titanium	Ti	Increases the altered surface hardness and reduces the surface cracks
Manganese	Mn	Defect-free surface with high surface microhardness and low SR
Hydroxyapatite	HA	Develops bioactive surfaces with enhanced wettability and high SR
Carbon nanotube	CNT	Low surface microcracks, process instability, SR and coated layer thickness
Zirconium	Zr	Improves the machined surface with low SR and TWR
Aluminium oxide	Al_2O_3	Nano-sized particles improve the surface finish and topography
Titanium dioxide	TiO_2	MRR reduces up to the optimum quantity of powder and then increases
Tungsten carbide	WC	High surface microhardness and wear resistance
Nickel	Ni	High abrasion resistance of the coated layer, MRR and TWR with low SR
Silver	Ag	Coated layer thickness increases with powder quantity
Copper	Cu	Surface quality and MRR increase with the optimal quantity of powder
Chromium	Cr	High microhardness and corrosion resistance with a low tool wear ratio
Molybdenum	Mo	Improves the machining efficiency and surface microhardness
Tungsten	W	Enhances surface integrity, MRR and the coated layer microhardness
Boron carbide	B_4C	Increases MRR and IEG and reduces TWR with good discharge distribution
Zinc	Zn	Enhances corrosion resistance and incorporates high MRR
Graphene	Graphene	Effectively augments the MRR and reduces SR and coated layer thickness
Titanium carbide	TiC	Forms a hard coated surface and assists in dispersion hardening with low SR
Boron nitride	BN	Increases MRR and develops a hard coated surface on the machined part
Niobium	Nb	Increases the adhesion strength of the coated layer and wear resistance
Boron oxide	B_2O_3	Increases MRR and surface microhardness of the coated layer
Tungsten disulphide	WS_2	Enhances the lubricity and wear resistance of the coated layer
Titanium nitride	TiN	Elevates microhardness, good corrosion, and wear resistance
Tantalum carbide	TaC	Strengthens surface microhardness and enhances process stability
Cobalt	Co	Improves adhesion strength, corrosion, and wear resistance

The PMEDM process trend is maximum in 2018, 2021 and 2022, with Gr, Al, Si and SiC powders being used most frequently during the last decade, as the bar chart representation is illustrated in Figure 5. The frequency on the y-axis represents the number of articles in a given year describing the particular powder usage in the EDM process.

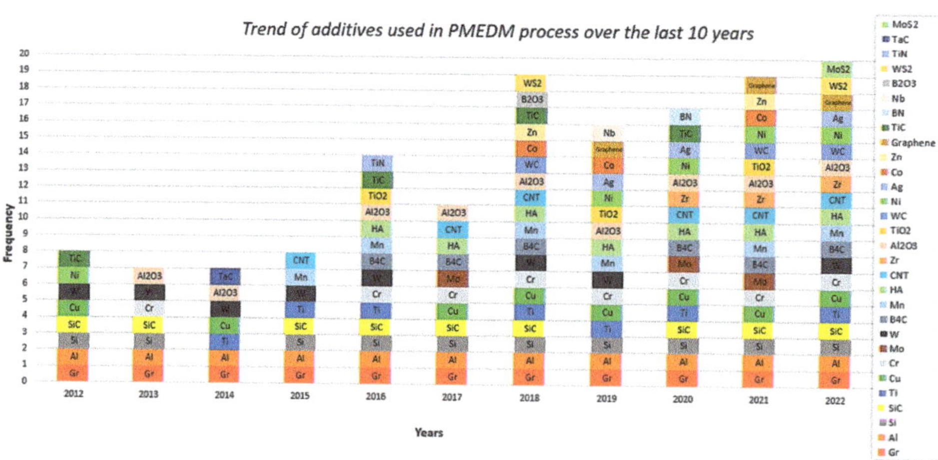

Figure 5. The trend of powder materials utilized for the PMEDM process (Extracted from the databases: Scopus and Web of Science).

4. Influential Process Parameters in the PMEDM

The process factors (electrical and non-electrical) significantly affect PMEDM performance. Some of the significant process parameters can be classified in the following way:

4.1. Dielectric Type

Kerosene, mineral oil, and pure distilled water are generally used as dielectrics in the PMEDM process, and their significant properties are tabulated in Table 2. Pure water possesses higher thermal conductivity and specific heat to withdraw heat from the machining area and ensure effective cooling [37]. During the machining process, kerosene builds carbides, and pure water creates oxides on the altered surface of the workpiece [38]. Carbides need higher thermal energy to dissolve than oxides, resulting in a better material removal rate, minimum tool wear rate for deionized water, and superior surface finish for kerosene as the dielectric.

Table 2. Significant properties of dielectrics used in the PMEDM process [39].

Dielectric Medium	Thermal Conductivity (W/m·K)	Specific Heat (J/kg·K)	Breakdown Strength (kV/mm)	Flashpoint (°C)
Kerosene	0.14	2100	24	37–65
Deionised water	0.62	4200	65–70	N/A
Mineral oil	0.13	1860	10–15	160
Silicon oil	0.15	1510	10–15	300

4.2. Peak Current

Peak current is one of the most significant process parameters that must be optimally allocated as it directly impacts PMEDM performance. The increase in the peak current produces more electrons and ions in the machining area, which develops higher pressure in the plasma channel and increases the material removal rate due to the boost up of discharge energy [40–42]. The peak current's rise may result from increased tool wear, surface quality degradation, recast layer thickness, deep craters, etc. [43,44]. However, the instant heating and quenching at a high-level pulse current results in the improved micro-hardness of the altered surface [45].

4.3. Pulse Duration

High pulse duration may lead to an unstable process and deteriorated surface quality by short-circuiting and insufficient debris exclusion from the machining region [43,46]. When the pulse duration exceeds the productive machining time, additional residues are formed and stick to the machined surface, causing an increase in the coated layer thickness [38,47]. The microhardness of the altered surface gradually improves with the increasing pulse duration due to the material transfer process [45].

4.4. Voltage Gap

An upper increase in the voltage gap during the machining process takes more time to bridge the ions and electrons within the discharge gap, augments the spark gap and lessens the energy density resulting in a lowering of surface irregularity and material removal rate of the altered surface [48–50]. A rise in gap voltage increases deposited layer thickness because of augmentation of the spark gap and reduction in coated layer thickness due to lowering the discharge column [51].

4.5. Polarity

The polarity designates the allocation of a cathode (−) and anode (+) onto the tool and workpiece. For achieving a better material erosion rate and lower tool wear rate, positive polarity on the workpiece with a shorter pulse duration and negative polarity with a longer pulse duration is preferable. The positive and negative polarities results were observed, as shown in Figure 6, using Scanning Electron Microscopy (SEM) after performing the EDM of a Ti alloy in which SiC was used as an additive in the dielectric [52]. The bulging response at the crater's center is examined with positive polarity causing the buildup of powder material on the machined surface [53]. In contrast, deep cavities were observed with negative polarity, resulting in a better erosion rate and a coarser surface finish [52,54].

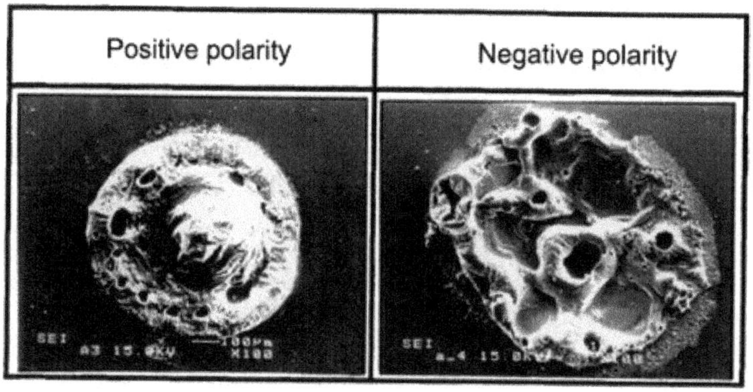

Figure 6. The outcome of polarity variation on the machined surface [52].

4.6. Duty Cycle

An optimum selection of the duty cycle percentage leads to reasonably low surface irregularity and an improved material erosion rate. The extended duty cycle causes process instability, insufficient flushing of accumulated debris, and electrode wear [48,50].

4.7. Electrode Classification

The electrode tool is an essential parameter of the PMEDM process which demands a low tool wear rate during the machining phase and a moderate tool wear rate during the coating phase of the PMEDM process. The PMEDM electrode mainly relies on its material, shape and size. The machining process develops a mirror shape of the electrode on the workpiece. The finished size of the workpiece is achieved by setting an undersized amount during the machining process to compensate for the electrode wear. The formula for the undersize amount can be given as follows:

$$\text{Undersize amount (one side)} = (\text{Finished size} - \text{Electrode size})/2$$

The undersized amount is incorporated in the electrode design, keeping the machining accuracy and projection area in view. The criteria for selecting the suitable electrode material for a specific output requirement depends on the properties tabulated in Table 3.

Table 3. Electrode classification and its properties [39,55].

Electrode Material	**Properties That Impact the Selection Criteria of EDM Electrode**							
	Density (g/cm^3)	**Melting Point (°C)**	**Thermal Conductivity (W/mK)**	**Electrical Resistivity (μΩ cm)**	**MRR [1]**	**TWR [2]**	**Cost**	**Fabrication**
Metals								
Copper	8.93	1083	385	1.7	High	Low	High	Easy
Tungsten	19.3	3370	163	5.65	Low	Low	High	Difficult
Brass	8.03	920	159	15.6	High	High	Low	Easy
Titanium	4.5	1668	17	178	High	Low	High	Difficult
Silver	10.49	961	419	1.55	High	Low	High	Medium
Aluminum	2.69	660	210	2.7	High	High	Low	Easy

Table 3. *Cont.*

Electrode Material	Properties That Impact the Selection Criteria of EDM Electrode							
	Density (g/cm^3)	Melting Point (°C)	Thermal Conductivity (W/mK)	Electrical Resistivity (μΩ cm)	MRR [1]	TWR [2]	Cost	Fabrication
Non-Metals								
Graphite	2.25	3650	25	6000	High	Low	High	Easy
Composites								
Cu-W	14.84	2250	220	3.83	Medium	Low	High	Medium
Cu-Gr	6.8	2550	250	4.36	High	Low	High	Easy
Ag-W	15.28	>980	160	3.45	Medium	Low	High	Difficult
W-C	15.7	2870	84.02	66.5	High	Low	High	Difficult
Te-Cu	2.69	660	210	2.7	Low	High	High	Difficult

[1] Material Removal Rate, [2] Tool Wear Rate.

4.8. Discharge Energy

Discharge energy is an essential factor that impacts surface topography, which refers to the profile shape and roughness of the altered surface. The energy of electric discharge is linked with other parameters of EDM and is a product of peak current, voltage, polarity and pulse duration [56]. Discharge energy determines the crater size and shape in micro-EDM applications [57]. This energy affects the coating thickness-to-roughness ratio, critically important when analyzing the coated layer on metallic biomaterials developed using the PMEDM process.

4.9. Surface Free Energy

The melting, vaporization and re-solidification of electrode material occur during the single discharge phase of the EDM process. During this process, energized debris gets deposited on the treated surface, a capacitive force is developed in the machining gap, and surface-free energy is acquired on the modified surface. A hydrophilic surface is formed due to the surface energy, shallow craters, and nanopores formation, improving biocompatibility and osseointegration at the bone-implant interface. The wettability of a coated surface is assessed based on the contact angle between the liquid and the solid surface. A surface is hydrophilic if the contact angle measurement is less than 90° and is desirable for bioimplant-coated surfaces [58].

5. Common Metallic Biomaterials Investigated Using the PMEDM Process

Metallic biomaterials account for 70% of bioimplant base material manufacturing mainly due to the advantages of mechanical strength, ductility, corrosion and wear resistance [59]. The surface modification of the biomaterials with biocompatible elements can develop a micron-level coating resulting in an effective bone-implant interface. Numerous physical and chemical coating methods are already established for surface modification of biomaterials, e.g., thermal spraying, laser cladding, sol-gel dip coating, electrochemical deposition, etc. [60]. Most prior techniques incorporate weak bonding between the coating film and substrate, lack process controllability, and contain a special apparatus with high-temperature devices. All require the fabrication of bioimplants before the coating process [61]. PMEDM is an emerging and ongoing research technique that can simultaneously perform the machining and coating process and possesses a high potential to serve as a cost-effective surface modification method to fulfil surface functional requirements, i.e., surface roughness, microhardness, corrosion resistance, biocompatibility, cracks-free and nano-porous surface for better osseointegration. Numerous studies have been commenced to explore the impact of multiple process factors indulged in powder-mixed EDM by investigating various combinations of work materials and powder additives. The research is still ongoing and requires a breakthrough in introducing the PMEDM process for the mass production of bio-implants in the marketplace.

5.1. Machined Surface of Biomaterial

The most frequently used biomaterials, for instance, Ti alloys, Mg alloys and 316L SS, possess exceptional mechanical properties but not satisfactory bio-function capabilities. During the PMEDM process, selecting appropriate electrodes, additive powder, and discharge energy significantly influence the substrate's altered surface to enhance the material erosion rate and minimize electrode wear rate and surface topography [13].

Bains et al. [62] analyzed the machined surface of Ti-6Al-4V by employing Copper (Cu) as an electrode and Hydroxyapatite (HA) as an additive to the dielectric medium. It was noticed that the peak current and powder concentration are the significant process factors for enhancing the material removal rate. The highest erosion rate of 7.648 mg/min was estimated for the current of 12 A. Bui et al. [63] assessed the machined surface of Ti alloy by utilizing Tungsten Carbide (WC) as an electrode tool and nano silver (Ag) particles as an additive to the dielectric medium. The high electrical and thermal conductivity of Ag powder particles results in a 185% increase in the material erosion rate and an 8.7% reduction in the electrode wear rate due to the WC's elevated melting point and hardness. Prakash et al. [64] assessed the machined surface of Ti-6Al-4V by using Cu as an electrode tool and Silicon (Si) powder as an additive to the dielectric medium. It was discovered that the peak current, pulse duration and powder concentration are the most influential process factors for the optimum machining performance measures. The maximum material erosion rate of 1.1872 mm^3/min and minimum electrode wear rate of 0.0750 mm^3/min were achieved with the optimum values of current intensity (9.55 A), pulse duration (255.95 s) and powder amount of (4 g/L). Moreover, Prakash et al. [64] stated further reduction in the electrode wear rate by using pure Ti as an electrode, as the addition of Si powder tends to form TiC on the surface of the electrode. Shabgard and Khosrozadeh [65] performed on the machined surface of Ti-6Al-4V by employing Cu as an electrode and carbon nanotube (CNT) as an additive to the dielectric medium. It was reported that initially, the material erosion rate was improved mainly due to the enlargement of the discharge gap as the CNT absorbed the machining heat due to its high electrical and thermal conductivity and afterwards both material removal rate and electrode wear rate were decreased.

Hourmand et al. [66] quantified the machined surface of the Al-Mg2Si composite matrix by employing Cu as an electrode and nano Al particles as an additive to the dielectric medium. It was observed that potential voltage, current intensity, and pulse duration are the most impactful factors affecting the material erosion rate. The highest value of material erosion rate of 0.0736 g/min was accomplished at the current intensity of 15 A, pulse duration of 200 μs, a potential voltage of 80 V and duty cycle of 0.55%. Santosh et al. [67] examined the machined surface of ZM21 Mg alloy by engaging Cu as an electrode and nano Gr particles as an additive to the dielectric medium. A valuable increase in the material erosion rate was noticed by adding the Gr particles with the current intensity of 11 A, pulse-on duration of 75 μs and pulse-off duration of 6 μs.

Santosh et al. [67] investigated the influence of CNT addition to the dielectric fluid on stainless steel machining outcomes via the PMEDM process. It was revealed that the pulse duration and powder concentration significantly influence the material erosion rate. Most importantly, the machining rate of 0.2503 g/min was measured with the add-on of 0.3 g/L CNT, pulse-on time of 10 μs and a constant current of 6 A. Lamichhane et al. [68] investigated the machined surface of 316L SS by adding the hydroxyapatite powder to the dielectric fluid. The study discovered an improved material erosion rate and electrode wear rate due to influential process factors such as current intensity, pulse duration and the powder amount. A 65.45% increase in material erosion rate was measured with a value of 19.01 g/min corresponding to the current intensity of 28 A, pulse-on duration of 60 μs and pulse-off duration of 120 μs. Moreover, a 53.07% growth in electrode wear rate was observed due to the higher value of powder quantity and pulse-off time. Banh et al. [69]

examined the performance measures of the Ti powder mixed EDM process for steel alloys utilizing two separate electrodes, namely Gr and Cu. This study ensured that the most impactful process factor is powder amount followed by current intensity, pulse-on time, polarity, workpiece material and pulse-off time, resulting in maximum material erosion and minimum electrode wear rates.

5.2. Coated Surface of Biomaterial

During the PMEDM process, the preference of the suitable tool and additive powder has a substantial effect on the modified or treated surface of the substrate to optimize the coating thickness, micro-hardness, surface roughness, adhesion strength, biocompatibility and resistance to wear and corrosion [13].

Bains et al. [62] studied the modified surface of Ti-6Al-4V by using Copper (Cu) as an electrode tool and Hydroxyapatite (HAp) as an additive to the dielectric medium. It was discovered that adding HAp resulted in a surface roughness reduction and a surge in wear resistance by 82%. The enhanced, modified surface of the Ti-alloy may result due to the placement of hard carbide and bio-ceramic layer. Prakash and Uddin [70] explored the influence of MWCNT on HAp-mixed dielectric fluid on β-type Ti with the Gr as electrode via the PMEDM process. The research outcome claims that the mixture of MWCNT/HAp produces enhanced biocompatibility and surface integrity due to the uniform spark propagation during the machining process. Abdul-Rani et al. [71] analyzed the treated surface of Ti-alloy in terms of surface roughness and morphology by employing tungsten and copper as a tool and nano-Al particles as an additive to the dielectric fluid. The addition of nano-Al powder by 3 g/L provided an exceptional increase in surface roughness by about 38.46% due to its high conductivity.

From the morphological point of view, micro-cracks and voids were improved on the modified surface. Prakash et al. [72] evaluated the modified surface of Ti-alloy by using pure Ti as an electrode and Si powder as an additive to the dielectric fluid. In this research, a biocompatible and durable modified surface of 1080 HV was obtained by developing a 15 μm white layer thickness containing carbides and oxides at the peak current value of 15 A, pulse duration of 50 μs and Si powder amount of 8 g/L. Furthermore, the friction coefficient of the modified surface improved by 60% due to the development of TiC and SiC on the treated surface. Similarly, Farooq et al. [73] assessed the modified surface of Ti-alloy by employing Cu as an electrode tool and Silicon (Si) powder as an additive to the dielectric medium. Due to the superior thermal conductivity of Si powder particles, the rapid sparking in the machining gap occurs, and the amount of additive contributed to the surface roughness and recast layer thickness of the modified surface. Moreover, the nano-porous surface of 50–200 nm was examined with the Si powder quantity of 5 g/L. Devgan and Sidhu [74] investigated the altered surface of the Ti-alloy by utilizing the HAp-added dielectric. The add-on of HAp developed a bio-ceramic layer with exceptional micro-hardness, crack-free surface, corrosion resistance and biocompatibility.

Hourmand et al. [66] estimated the modified surface of Mg-alloy by employing Mg-Ca as an electrode and HAp particles as an additive to the dielectric medium. It was perceived that the increased concentration of HAp enhances the micro-hardness and wear resistance, whereas it reduces the surface roughness. The biocompatibility of the treated surface was achieved due to the formation of oxides, and the deposition layer thickness increased as the amount of HAp was enhanced. Razak et al. [75] assessed the modified surface of Mg-alloy by engaging Cu as an electrode tool and Zinc (Zn) powder particles as an additive to the dielectric medium. The high conductivity of Zn powder maintains the spark functioning and machining stability. Its optimum addition to the dielectric efficiently reduces surface roughness and wear resistance. Rout et al. [76] analyzed the treated surface of the Mg-alloy by using three electrodes of brass, copper and stainless steel and Al powder particles as an additive to the dielectric medium. It was revealed that the brass and copper electrodes and suspended Al powder enhance surface roughness. The performance of brass and copper electrodes was better than stainless steel electrodes because of their higher conductivity and stability in the machining process.

Lamichhane et al. [68] explored the impact of HAp addition to the dielectric medium to alter the surface of stainless steel using the PMEDM process. It was discovered that mixing HAp with the dielectric improves the surface roughness due to the scouring effect. A biocompatible surface was formed onto the modified surface because of the development of an oxide layer. Mohan et al. [77] investigated the modified surface of 316L SS by using tungsten as an electrode and the TiO_2 powder particles as an additive to the dielectric fluid. The study revealed that forming hard oxide and carbide layers onto the treated surface provides hardness and wear resistance, respectively. G. Singh et al. [78] evaluated the modified surface of 316L SS by utilizing Cu as an electrode and HAp powder particles as an additive to the dielectric fluid. In this research, the optimum add-on of HAp powder produces a hard and biocompatible oxide layer on the machine surface resulting in the enhanced micro-hardness and bioactivity of the developed bioimplants. Bhaumik and Maity [79] assessed the modified surface of 316L SS by engaging W-C as an electrode tool and SiC powder particles as an additive to the dielectric medium. The research revealed that adding SiC forms a hard carbide layer that enhances surface hardness, wear resistance and surface roughness.

6. Methodologies for Optimization of PMEDM Process

The main objective of employing optimization methods in the PMEDM process is to find the values of the influential process parameters for optimising performance responses. The accomplishment of optimum settings for the machining parameters of the PMEDM process is unlike the EDM process due to the involvement of powder additives. Numerous researchers have employed tools like the design of experiment (DOE) methods, analysis of variance (ANOVA) and numerical techniques to discover the optimal values for the PMEDM process and performance parameters. Single and multiple objective optimization techniques estimate the best solution for a specific criterion.

6.1. Taguchi Method

Taguchi is a robust statistical tool to plan the experimentation for the best values of process factors. Assarzadeh and Ghoreishi [48] used the Taguchi method to obtain the optimum process parameters by employing boron carbide-mixed electric discharge machining on the surface of titanium alloy. It was revealed via ANOVA that current intensity, pulse duration and powder quantity are the most influential factors contributing to the machining performance measures, i.e., material erosion rate and surface roughness. Kansal et al. [80] applied the Taguchi method to attain the optimum process factors by engaging graphite powder particles in the kerosene oil dielectric for performing the rough machining process on the surface of die steel (AISI D2). The experimental investigative results reveal that the suitable addition of graphite particles to the dielectric enhances the material erosion rate and reduces electrode wear and surface irregularity. Ishfaq et al. [81] employed graphene nanoparticles in the dielectric medium for sustainable EDM of Ti-alloy by using three different electrodes Al, Cu and Br. The study carried out the planning for the design of experimentation using Taguchi's method and analyzed the experimental results using statistical tools. Bains et al. [62] engaged Hydroxyapatite (HAp) powder particles in the dielectric medium to machinate Ti-6Al-4V. Experiments were executed based on Taguchi design with an orthogonal array of L16 to consider process parameters, i.e., current intensity, pulse-on and off time, electrode inner diameter and dielectric type concerning outcome responses in the form of surface irregularity and material erosion rate. The machining rate was statistically significant ($p < 0.05$) by the factors of hole diameter, peak current, dielectric fluid, and the combination of current and dielectric. SR was statistically significant by current intensity, pulse duration and dielectric medium.

6.2. Response Surface Methodology (RSM)

RSM is a mathematical and statistical method used for modelling and analysing a process for an optimized response measure controlled by variable factors. Garg and Ojha [82] studied the machined surface of EN-8 steel by adding chromium powder particles as an additive to the dielectric. The RSM method has been used to plan and analyse the experimentation for process variables and consider PMEDM performance in improved material removal rate and surface finish. Aliyu et al. [83] investigated the Hydroxyapatite Deposition Rate (HDR) and Surface Roughness (SR) during the electro-discharge coating of Bulk Metallic Glass (BMG). Using a D-optimum customizable design approach, RSM was employed to generate the models and optimize the input factors. Optimum process parameters setting is essential in enhancing product performance and contributing to the commercial industries by lowering the component's production time and expense.

6.3. Analytic Hierarchy Process (AHP)

AHP is a technique used to organise and indicate process variables' importance on output performance with an analytical approach. Bhattacharya et al. [84] executed a multi-objective optimization of material erosion rate, electrode wear rate and surface roughness during PMEDM of High Carbon High Chromium (HCHC), Hot Die Steel (HDS) and EN31 using the AHP approach.

6.4. Gray Relational Analysis (GRA)

GRA is an approach used to obtain optimum conditions of input variables to find the best-suited outcome with insufficient information. S. Singh and Yeh [33] and Talla et al. [85] employed the GRA technique to assess the efficacy of optimizing the variable performance measures of PMEDM for aluminum matrix composites with inadequate data.

6.5. Non-Sorted Genetic Algorithm (NSGA-II)

NSGA-II is an advanced evolutionary algorithm used for multi-objective optimization. Al-Amin et al. [86,87] investigated the effects of process variables (current intensity, pulse duration, powder amount, and duty cycle) on machining performance (material erosion rate) and surface features (surface irregularity and recast layer thickness) using multi-objective algorithms for optimal processing of MWCNT and HAp added ED machined surface of 316L SS. NSGA-II was used as a multi-objective optimisation tool and Pareto frontiers for optimal solution sets using MATLAB software. The LINMAP method selected the finest solution set outcomes based on defined objective functions. Padhee et al. [88] carried out concurrent optimization of several targets using NSGA and provided multiple outcomes to the manufacturer for sorting the optimal combination from the approachable resources.

6.6. Hybrid Methods

The multi-objective optimization technique is used to optimize two or more opposing objectives. The concurrent employment of the single and multiple optimisation approaches leads to the hybrid response.

Tripathy and Tripathy [89] performed the PMEDM process on the machined surface of H-11 die steel using chromium-mixed dielectric and copper electrodes. The Taguchi method, combined with TOPSIS and GRA, has been implemented to estimate the usefulness of enhancing numerous performance characteristics.

Prakash et al. [90] estimated the optimal settings of PMEDM process factors to produce the biocompatible surface on β-phase Ti alloy by employing NSGA-II coupled with Taguchi-based RSM for the first time. The optimal condition was achieved with the desired requirement of high surface microhardness and low surface roughness.

6.7. Numerical Modeling

An emerging trend among researchers is heading towards the numerical modelling of the PMEDM process to simulate the temperature distribution and surface features onto the machined surface. Kansal et al. [91] established an axisymmetric 2D thermal model for the PMEDM process using the Finite Element Method (FEM) technique. The study estimated the temperature distribution on the substrate and measured the material erosion rate from the temperature profiles. Bhattacharya et al. [92] achieved a 3D FEM model to estimate the residual thermal stresses incorporated during the PMEDM process. Additionally, a mathematical model was developed by Bhattacharya and Batish [93] for predicting the crater dimensions and volume removed during the PMEDM process. Tan and Yeo [94] simulated a 3D FEM model for surface integrity features for obtaining enhanced surface roughness and reduced recast layer thickness using crater theories.

Wandra [95] numerically simulated the material removal mechanism using the finite element analysis of the PMEDM process on the Ti-alloy. The simulation results found that the surface finish and crater geometry on the machined surface during powder-mixed EDM improved much compared to the EDM process. Tumer et al. [96] investigated the mechanical properties of the graphene/CNT-coated Ti-6Al-4V and 316L SS using the finite element analysis approach.The loading impact on the femur, first screw and the plate with coated and uncoated conditions. Jampana [97] examined the thermal analysis on powder-mixed EDM of stainless steel 630 using finite element and artificial neural network techniques for obtaining optimum performance parameters regarding material erosion rate and surface irregularity.

7. Applicability of the PMEDM Process

EDM can precisely manufacture critical components (i.e., impeller, fuel system, engine blocks, piston heads, etc.) for the aerospace and automobile industries. Still, on the contrary, it lacks the surface integrity features that can be potentially achieved using the PMEDM process. With enhanced dimensional accuracy, PMEDM introduces exceptional surface quality, biocompatibility and bioactivity, emerging to be an innovative advanced machining option for the manufacturing industries of bio-implants and surgical instruments. The potential application areas of PMEDM can be classified in the following way:

7.1. Rough Machining

The process parameter values of high current intensity and pulse duration mainly cause the rough machined surface [98]. In this case, the problem arises with the rough machined surface in the form of inadequate flushing of debris and a high tool wear rate [17]. These issues can be addressed efficiently by considering the optimal values of process factors. Mai et al. [10] employed CNTs in the dielectric medium using PMEDM of NAK-80 die steel for evaluating the rough machining factors. It was revealed that long pulse duration and high peak current significantly increase the machining rate.

7.2. Fine Machining

Finish machining is one of the significant fields of application in the PMEDM process. The ability to produce a finely finished surface makes PMEDM a distinguished machining option that avoids finishing operations costs. A near-mirror surface finish can be achieved using the PMEDM process by increasing tool size and low discharge energy settings. Mohri et al. [99] and Pecas and Henriques [100] observed a high-quality surface finish during the electric discharge machining of H13 steel using Si-powder-mixed dielectric, as shown in Figure 7, and the finish quality is further improved by increasing the machining time. A further study examined a mirror-like finish on the machined surface of SKH-54 by adding Al powder particles as an additive to the dielectric medium.

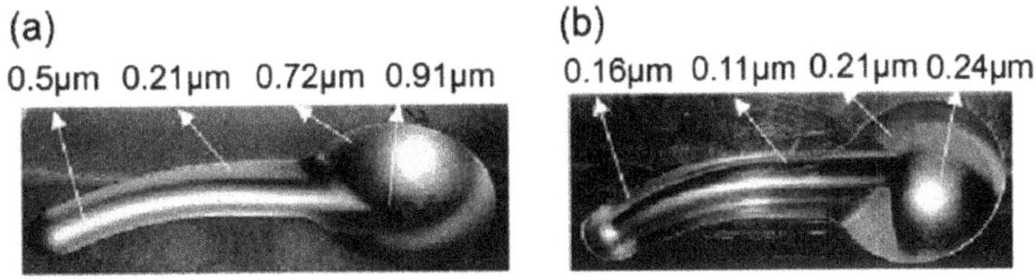

Figure 7. Mold cavity (**a**) without Si powder (**b**) with Si powder [100].

7.3. Micromachining

Technological advancements in micro-electro-mechanical systems (MEMS) initiate the demand for micromachining processes. The micro-mechanical accessories (e.g., mini pumps, engines, and robots) and application areas (e.g., drilling of micro-holes for cooling purposes on turbine blades) require precise micromachining procedures. Kibria and Bhattacharyya [38,101,102] employed the μ-PMEDM approach for drilling holes on the Ti-alloy surface using boron carbide mixed with kerosene and pure water as a dielectric medium. Chow et al. [18] revealed a potential application of the PMEDM process to fabricate a micro-heat scatter fin on the surface of Ti-6Al-4V, as displayed in Figure 8. The optimum discharge conditions can be classified as peak current = 0.1 A, pulse-on time = 10 μs, workpiece size = 0.8 × 0.8 × 0.6 mm, copper as tool electrode and kerosene as dielectric fluid.

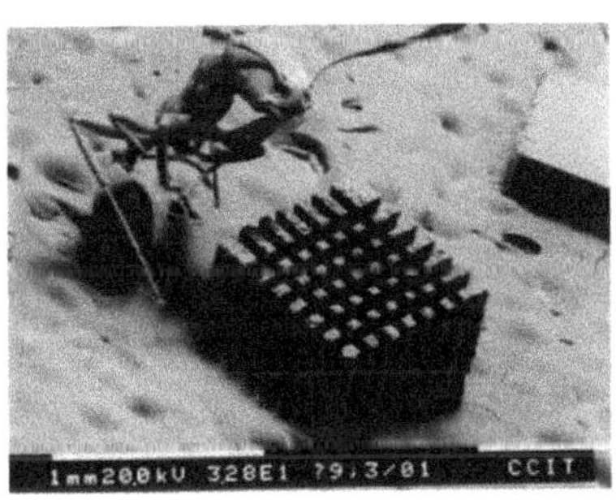

Figure 8. SEM view of micro-slits fabricated on the surface of Ti-alloy using SiC-mixed EDM process [20].

7.4. Surface Modification

Inaccuracies arising from high tool wear avoid the utilization of electro-discharge coating (EDC) as a powder deposition approach for surface coating and modification. PMEDM is an arising technique for simultaneously machining and alloy coating onto the modified surface with enhanced precision. Chen et al. [103] enhanced biocompatibility by forming TiO on the modified surface of pure Ti using PMEDM with Ti-added pure water. Zain et al. [40] accomplished exceptional microhardness on the modified surface of SUS 304 stainless steel using tantalum carbide powder as an additive to the dielectric. Bhattacharya et al. [104] evaluated the influence of different tool and powder combinations on the micro-hardness of the modified surface. The study found that the merger of the W-Cu tool and W powder particles to the dielectric develops the hardest modified surface compared to the surfaces achieved using Gr and Si powder-mixed dielectrics.

7.5. Machining of Non-Conductive Materials

The machining of non-conductive materials in the PMEDM process is yet to be explored. Kucukturk & Cogun [105] performed non-conductive ceramics via PMEDM by mixing Gr powder in the dielectric medium. The machined surface was coated with a conductive layer to ensure the sparking.

8. Future Research Directions of the PMEDM Process

In the past ten years, several researchers have published review articles on the potentiality of the electric discharge machining and coating process for surface modification and highlighted research directions yet to be explored. The ISI-indexed review articles based on journal citation reports (JCR) have been tabulated chronologically in Table 4.

Table 4. Review articles on the PMEDM process over the past decade.

Review Article (Year/JIF [1]/Quartile)	Journal/Publisher	Future Research Directions
[106] An insight on Powder-Mixed Electric Discharge Machining: A state-of-the-art review (2022/2.75/Q2)	Journal of Engineering Manufacture/SAGE	1. PMEDM process associated with composite coated electrodes. 2. PMEDM process coupled with laser and ultrasonic machining. 3. Employment of shape memory alloys in the PMEDM process. 4. Deep learning algorithms for the PMEDM process.
[107] Investigation of Coatings, Corrosion and Wear Characteristics of Machined Biomaterials through Hydroxyapatite Mixed-EDM Process: A Review (2021/3.62/Q2)	Materials/MDPI	1. Uniform thin coating formation with homogeneous alloying. 2. Uniform distribution of HAp-based oxide and carbide on the deposited coat and its phase analysis. 3. HAp doped with reinforced additives for proper distribution and enhancing mechanical properties of biomaterials. 4. Measuring residual stress and wettability of the coating.
[108] Transition from EDM to PMEDM—Impact of suspended particulates in the dielectric on Ti6Al4V and other distinct material surfaces: A review (2021/5.01/Q2)	Journal of Manufacturing Processes/ ELSEVIER	1. Evaluating the Recast Layer (RL) metallurgical behavior (phase distribution and grain size/refinement), bond strength (coherence with the substrate), and homogeneity of the various surface compounds over a specific area. 2. The optimum powder concentration range by reducing the particles' size from the micro to the nanoscale. 3. Enhancing uniform spread of coated layer and dispersion of the powder particles to avoid agglomeration. 4. The effect of the shape and size of the powder particles and tool electrode on the outcome responses. 5. Factors affecting material melting/vaporization / re-solidification and alteration in the craters' morphology (size and shape). 6. Multiple powders in the dielectric medium to improve the tribological and bio-adaptive characteristics using PMEDM. 7. Develop realistic practical models for improving the PMEDM process considering its stochastic nature.
[109] Bio-ceramic coatings adhesion and roughness of biomaterials through PM-EDM: a comprehensive review (2020/4.61/Q2)	Materials and Manufacturing Processes/ TAYLOR & FRANCIS INC	1. Assuring uniform bio-ceramic coating for improving biocompatibility. 2. Suitable distribution of nano-porosities on the modified surface. 3. Improvement by effectively adding surfactant with HAp for uniform dispersion in dielectric fluid. 4. Optimal discharge energy for enhanced adhesion and surface roughness.
[13] Assessment of PM-EDM cycle factors influence on machining responses and surface properties of biomaterials: A comprehensive review. (2020/3.31/Q2)	Precision Engineering/ELSEVIER	1. Quantification of nano-porous surface and its porosity distribution on the machined surface. 2. Uniform dispersion of additives during the machining process. 3. Formulation for selection of suitable additives for biomaterials. 4. Controlling the presence of amorphous elements during the coating process.

Table 4. *Cont.*

Review Article (Year/JIF [1]/Quartile)	Journal/Publisher	Future Research Directions
[58] Powder mixed-EDM for potential biomedical applications: A critical review. (2020/4.61/Q2)	Materials and Manufacturing Processes/ TAYLOR & FRANCIS INC	1. Developing uniform thickness of the coating. 2. Formulating optimum process factors for enhancing surface characteristics and mechanical properties. 3. Development of the anti-bacterial coating as per standardization. 4. Minimizing toxic and corrosive active elements and compounds during the coating process.
[110] Recent Advances and Perceptive Insights into Powder-Mixed Dielectric Fluid of EDM (2020/2.89/Q2)	Micromachines/ MDPI	1. Exploration of PMEDM mechanism to find an optimum relationship among the machining parameters, powder properties, machined and modified surface characteristics. 2. Environmental health issues related to using a dielectric fluid with additives. 3. Machining of non-conductive materials. 4. Reduction of machining time in case of precisely accurate parts.
[39] A systematic review on powder mixed electrical discharge machining. (2019/0.46/Q2)	Heliyon/ ELSEVIER	1. Proper pumping mechanism and dielectric selection for smooth powder particle flow. 2. Optimum concentration of powder particles for surface modification. 3. Agglomeration and settlement of powder particles at the bottom of the tank. 4. Separation of powder particles and debris for reusability.
[14] State of the art in powder-mixed electric discharge machining: A review (2017/2.61/Q2)	Journal of Engineering Manufacture/SAGE	1. An efficient mechanism and appropriate dielectric for flushing powder particles. 2. Collection of dense powder particles at the tank base. 3. The required amount of powder particles for surface modification. 4. Eco-friendly process. 5. System for separation of debris from the powder particles.
[111] A Review of Additive Mixed-Electric Discharge Machining: Current Status and Future Perspectives for Surface Modification of Biomedical Implants (2017/1.72/Q3)	Advances in Materials Science and Engineering/ Hindawi	1. Controlling and measuring the actual thickness of the coated layer. 2. Optimum relationship between powder concentration and layer thickness size for surface coating. 3. Material deposition monitoring during the machining process. 4. Uniformity of coated layer to provide implant-bone bonding. 5. Formulation of optimum process parameters for all powders.
[112] State of the art in powder mixed dielectric for EDM applications (2016/3.31/Q2)	Precision Engineering/ELSEVIER	1. Effect of particle shape on performance parameters. 2. Comparison between various powder materials, sizes, and concentrations for finding the powder-specific application of PMEDM. 3. Relationship between powder and concentration with the influence of vibration frequency and amplitude on PMEDM performance.
[113] Current trends in electric discharge machining using micro and nano powder materials—A Review (2015/26.94/Q1)	Materials Today/ELSEVIER	1. Effect of particle shape on the performance parameters, for instance, machining efficiency, surface finish and tribological properties, are yet to be explored in the PMEDM process. 2. Impact of particle shape while considering micro/nano levelled powder particles.

[1] Journal Impact Factor.

9. Challenges to Commercialize PMEDM Process

The PMEDM process is still open-ended and requires a technological breakthrough to enter the production market. Numerous researchers have claimed that superior surface quality and enhanced productivity can be achieved with the PMEDM process, but still, several challenges impede the embarkation of PMEDM industrialization.

9.1. Cost Effectiveness

The accumulation of powder particles at the tank's base adds cost and alters the powder concentration during the machining process [112]. The homogeneity of the powder particles in the dielectric medium can be explored by adding an effective surfactant to the powder-mixed dielectric, and ultrasonic vibration may prevent the settlement of powder particles. An effective PMEDM circulation system is desired that launches a commercially cost-efficient solution to restrain the powder agglomeration and deposition at the base of the machining tank [112]. Figure 9 represents the illustration of the PMEDM circulation setup. The mixing of debris with the powder material may primarily affect the outcome. Magnetic filters can be introduced to separate the debris in the filter system to avoid mixing with the dielectric and powders.

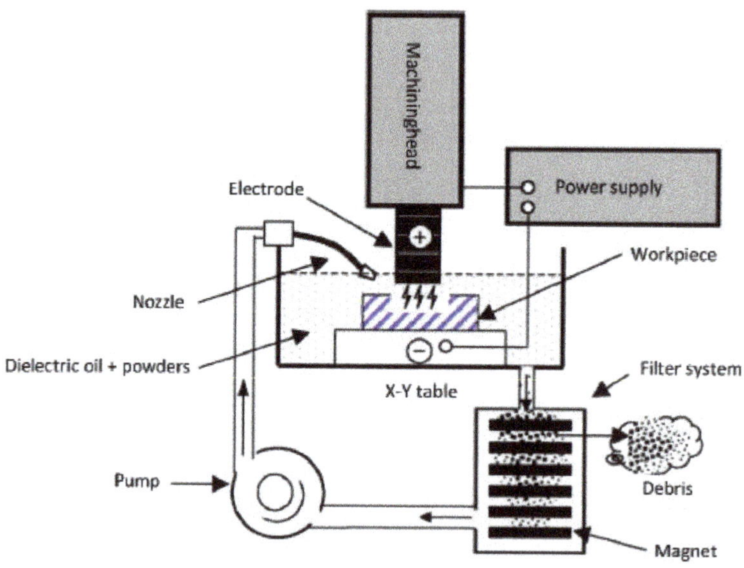

Figure 9. Representation of PMEDM Circulation Setup [112].

9.2. Longevity of the Deposited Layer

The researchers do not claim the life span of the coated layer deposited by the PMEDM process. However, multiple research studies have evaluated the adhesion strength, micro-hardness, and degradability of the coated layer onto the machined surface [112].

9.3. Reusability of the Powders

During the PMEDM process, the chemical transformation of the powders (micro and nano size) can occur, which restricts their re-use and may require procuring the powders in large quantities for experimentation. During the coating process, a minimal amount of powder is utilized on the substrate, and a significant amount is flushed out. Due to the stochasticity of the process, the justification for powder consumption and determination of powder concentration for a specific application is a challenge for the commercialization of the PMEDM process.

9.4. Machining Time

Industrial manufacturing processes are generally for lengthy durations. The experimental studies of the PMEDM process usually require a short machining time. For industrial applications, it would be a real challenge to introduce this process for the long period of machining and coating work.

9.5. In Situ Monitoring of the PMEDM Process

The addition of powders in the submerged machining process may obstruct the visibility of the ongoing process. With the technological advancements in manufacturing processes, advanced industries are pursuing on-stage monitoring to ensure the dimensional accuracy of manufactured parts. In the PMEDM process, adding the powder to the dielectric fluid makes it challenging to monitor and visualize the outcome during the machining and coating process.

9.6. Health, Safety and Environmental Considerations

The social responsibility of PMEDM users demands deliberation on health and environmental safety measures. The operator's exposure to the fumes generated from powder mixed dielectric, the explosibility of powder materials, and discarding dielectric are critical factors that require specific working procedures. International standards can be developed and implemented after certification from the regulatory authorities.

10. Limitation

This review study has a potential limitation that covers only the general understanding of the PMEDM process and its effectiveness. The transition from EDM to PMEDM discovered an emerging technological approach to simultaneously machine and coat materials, i.e., still at the early research stage. This technology got the attention of researchers to functionalize it on biomaterials for easy, inexpensive and standard compliance manufacturing of bioimplants. Therefore, this article also demonstrates the utilization of the PMEDM process for the machining and coating metallic biomaterials.

11. Summary

From the deliberations, the PMEDM method is recommended as an efficient emerging manufacturing process that can perform concurrent coating and machining of biomaterials. The performance parameters of the PMEDM process, such as MRR, TWR, surface roughness, recast layer thickness, bonding/adhesion strength, microhardness, biocompatibility, and resistance to wear and corrosion depend significantly on the correlated process factors due to its stochastic behavioral mechanism. Consequently, appropriate formulation of the process factors and the additive powders are required to achieve the significant performance measures of this process before its industrialization.

This article reviews the PMEDM process, applicability, trend, and futuristic perspective based on the current research study. It highlights the mechanism of the PMEDM process, the impact of powder attributes and EDM process parameters on the performance parameters. It also reviews the influence of tool electrodes and additive powders on the machined and modified surface of commonly used biomaterials. Furthermore, the study represents the most used methods for optimising the PMEDM process, futuristic research directions, challenges, and research trends over the past decade.

Author Contributions: Conceptualization, Visualization, Writing—original draft, Manuscript preparation, I.A.G.; Supervision, Review and Editing, Project management, Funding acquisition, A.M.A.-R.; Review and Editing, Resources, M.A.-A.; Review and Editing, E.G. All authors have read and agreed to the published version of the manuscript.

Funding: This research study was supported by the Malaysian Ministry of Higher Education Fundamental Research Grant Scheme (FRGS/1/2020/TK0/UTP/02/39) and YUTP (Cost center: 015LC0-385).

Institutional Review Board Statement: Not applicable.

Informed Consent Statement: Not applicable.

Data Availability Statement: Not applicable.

Acknowledgments: On behalf of the institution, the authors articulate their profound appreciation to the Malaysian Ministry of Higher Education Fundamental Research Grant Scheme (FRGS/1/2020/TK0/-UTP/02/39) and YUTP (Cost center: 015LC0-385) for providing the research work opportunity. The authors would also like to thank the UTP Mechanical Engineering Department for their assistance.

Conflicts of Interest: The authors declare no conflict of interest.

References

1. Schumacher, B.; Krampitz, R.; Kruth, J.-P. Historical Phases of EDM Development Driven by the Dual Influence of "Market Pull" and "Science Push". *Procedia CIRP* **2013**, *6*, 5–12. [CrossRef]
2. Luis, C.; Puertas, I.; Villa, G. Material removal rate and electrode wear study on the EDM of silicon carbide. *J. Mater. Process. Technol.* **2005**, *164–165*, 889–896. [CrossRef]
3. Mahajan, A.; Sidhu, S.S. Enhancing biocompatibility of Co-Cr alloy implants via electrical discharge process. *Mater. Technol.* **2018**, *33*, 524–531. [CrossRef]
4. Ho, K.; Newman, S. State of the art electrical discharge machining (EDM). *Int. J. Mach. Tools Manuf.* **2003**, *43*, 1287–1300. [CrossRef]
5. Abdulkareem, S.; Khan, A.A.; Konneh, M. Cooling Effect on Electrode and Process Parameters in EDM. *Mater. Manuf. Process.* **2010**, *25*, 462–466. [CrossRef]
6. Abu Qudeiri, J.E.; Saleh, A.; Ziout, A.; Mourad, A.-H.I.; Abidi, M.H.; Elkaseer, A. Advanced Electric Discharge Machining of Stainless Steels: Assessment of the State of the Art, Gaps and Future Prospect. *Materials* **2019**, *12*, 907. [CrossRef]
7. Razak, M.A.; Abdul-Rani, A.M.; Nanimina, A.M. Improving EDM Efficiency with Silicon Carbide Powder-Mixed Dielectric Fluid. *Int. J. Mater. Mech. Manuf.* **2015**, *3*, 40–43. [CrossRef]
8. Sonawane, H.A.; Pawade, R.S. Effects of powder mixed dielectric on electro discharge machining (PMEDM) of HSS tool steel. *Int. J. Mechatronics Manuf. Syst.* **2012**, *5*, 431. [CrossRef]
9. Puertas, I.; Luis, C. A study on the machining parameters optimisation of electrical discharge machining. *J. Mater. Process. Technol.* **2003**, *143–144*, 521–526. [CrossRef]
10. Mai, C.; Hocheng, H.; Huang, S. Advantages of carbon nanotubes in electrical discharge machining. *Int. J. Adv. Manuf. Technol.* **2011**, *59*, 111–117. [CrossRef]
11. Erden, A.; Bilgin, S. Role of Impurities in Electric Discharge Machining. In Proceedings of the Twenty-First International Machine Tool Design and Research Conference, Swansea, UK, 8–12 September 1980; pp. 345–350. [CrossRef]
12. Silicon Powder Additives Proven Safe and Effective in EDM | Modern Machine Shop. Available online: https://www.mmsonline.com/articles/silicon-powder-additives-proven-safe-and-effective-in-edm (accessed on 11 November 2022).
13. Amin, A.; Abdul-Rani, A.M.; Danish, M.; Thompson, H.M.; Aliyu, A.A.A.; Hastuty, S.; Zohura, F.T.; Bryant, M.G.; Rubaiee, S.; Rao, T. Assessment of PM-EDM cycle factors influence on machining responses and surface properties of biomaterials: A comprehensive review. *Precis. Eng.* **2020**, *66*, 531–549. [CrossRef]
14. Talla, G.; Gangopadhayay, S.; Biswas, C. State of the art in powder-mixed electric discharge machining: A review. *Proc. Inst. Mech. Eng. Part B J. Eng. Manuf.* **2016**, *231*, 2511–2526. [CrossRef]
15. Kumar, H. Development of mirror like surface characteristics using nano powder mixed electric discharge machining (NPMEDM). *Int. J. Adv. Manuf. Technol.* **2014**, *76*, 105–113. [CrossRef]
16. Prihandana, G.S.; Mahardika, M.; Hamdi, M.; Wong, Y.S.; Mitsui, K. Accuracy improvement in nanographite powder-suspended dielectric fluid for micro-electrical discharge machining processes. *Int. J. Adv. Manuf. Technol.* **2011**, *56*, 143–149. [CrossRef]
17. Zhao, W.; Meng, Q.; Wang, Z. The application of research on powder mixed EDM in rough machining. *J. Mater. Process. Technol.* **2002**, *129*, 30–33. [CrossRef]
18. Chow, H.-M.; Yang, L.-D.; Lin, C.-T.; Chen, Y.-F. The use of SiC powder in water as dielectric for micro-slit EDM machining. *J. Mater. Process. Technol.* **2008**, *195*, 160–170. [CrossRef]
19. Kung, K.-Y.; Horng, J.-T.; Chiang, K.-T. Material removal rate and electrode wear ratio study on the powder mixed electrical discharge machining of cobalt-bonded tungsten carbide. *Int. J. Adv. Manuf. Technol.* **2007**, *40*, 95–104. [CrossRef]
20. Ekmekci, B.; Ersöz, Y. How Suspended Particles Affect Surface Morphology in Powder Mixed Electrical Discharge Machining (PMEDM). *Met. Mater. Trans. B* **2012**, *43*, 1138–1148. [CrossRef]
21. Peng, H.; Yu, D.; Zhang, X.; Wang, S.; Wen, Y. Fabrication of hollow nickel micro-spheres with high degree of hollowness by silicon powder-mixed spark erosion. *Int. J. Mach. Tools Manuf.* **2014**, *85*, 131–134. [CrossRef]
22. Chow, H.-M.; Yan, B.-H.; Huang, F.-Y.; Hung, J.-C. Study of added powder in kerosene for the micro-slit machining of titanium alloy using electro-discharge machining. *J. Mater. Process. Technol.* **2000**, *101*, 95–103. [CrossRef]
23. Janmanee, P.; Muttamara, A. Surface modification of tungsten carbide by electrical discharge coating (EDC) using a titanium powder suspension. *Appl. Surf. Sci.* **2012**, *258*, 7255–7265. [CrossRef]

24. Prakash, C.; Kansal, H.K.; Pabla, B.; Puri, S.; Aggarwal, A. Electric discharge machining—A potential choice for surface modification of metallic implants for orthopedic applications: A review. *Proc. Inst. Mech. Eng. Part B J. Eng. Manuf.* **2015**, *230*, 331–353. [CrossRef]
25. Wong, Y.; Lim, L.; Rahuman, I.; Tee, W. Near-mirror-finish phenomenon in EDM using powder-mixed dielectric. *J. Mater. Process. Technol.* **1998**, *79*, 30–40. [CrossRef]
26. Tzeng, Y.-F.; Lee, C.-Y. Effects of Powder Characteristics on Electrodischarge Machining Efficiency. *Int. J. Adv. Manuf. Technol.* **2001**, *17*, 586–592. [CrossRef]
27. Yih-Fong, T.; Fu-Chen, C. Investigation into some surface characteristics of electrical discharge machined SKD-11 using powder-suspension dielectric oil. *J. Mater. Process. Technol.* **2005**, *170*, 385–391. [CrossRef]
28. Jeswani, M. Effect of the addition of graphite powder to kerosene used as the dielectric fluid in electrical discharge machining. *Wear* **1981**, *70*, 133–139. [CrossRef]
29. Prihandana, G.S.; Mahardika, M.; Hamdi, M.; Wong, Y.S.; Miki, N.; Mitsui, K. Study of workpiece vibration in powder-suspended dielectric fluid in micro-EDM processes. *Int. J. Precis. Eng. Manuf.* **2013**, *14*, 1817–1822. [CrossRef]
30. Marashi, H.; Sarhan, A.A.; Hamdi, M. Employing Ti nano-powder dielectric to enhance surface characteristics in electrical discharge machining of AISI D2 steel. *Appl. Surf. Sci.* **2015**, *357*, 892–907. [CrossRef]
31. Kumar, S.; Batra, U. Surface modification of die steel materials by EDM method using tungsten powder-mixed dielectric. *J. Manuf. Process.* **2012**, *14*, 35–40. [CrossRef]
32. Kumar, A.; Maheshwari, S.; Sharma, C.; Beri, N. Analysis of Machining Characteristics in Additive Mixed Electric Discharge Machining of Nickel-Based Super Alloy Inconel 718. *Mater. Manuf. Process.* **2011**, *26*, 1011–1018. [CrossRef]
33. Singh, S.; Yeh, M.F. Optimization of Abrasive Powder Mixed EDM of Aluminum Matrix Composites with Multiple Responses Using Gray Relational Analysis. *J. Mater. Eng. Perform.* **2011**, *21*, 481–491. [CrossRef]
34. KSyed, H.; Kuppan, P. Studies on recast-layer in EDM using aluminium powder mixed distilled water dielectric fluid. *Int. J. Eng. Technol.* **2013**, *5*, 1775–1780.
35. Kumar, H.; Davim, J.P. Role of Powder in the Machining of Al-10%Sic_p Metal Matrix Composites by Powder Mixed Electric Discharge Machining. *J. Compos. Mater.* **2010**, *45*, 133–151. [CrossRef]
36. Jahan, M.P.; Rahman, M.; Wong, Y.S. Study on the nano-powder-mixed sinking and milling micro-EDM of WC-Co. *Int. J. Adv. Manuf. Technol.* **2010**, *53*, 167–180. [CrossRef]
37. Lin, Y.C.; Yan, B.H.; Chang, Y.S. Machining characteristics of titanium alloy (Ti–6Al–4V) using a combination process of EDM with USM. *J. Mater. Process. Technol.* **2000**, *104*, 171–177. [CrossRef]
38. Kibria, G.; Sarkar, B.R.; Pradhan, B.B.; Bhattacharyya, B. Comparative study of different dielectrics for micro-EDM performance during microhole machining of Ti-6Al-4V alloy. *Int. J. Adv. Manuf. Technol.* **2009**, *48*, 557–570. [CrossRef]
39. Joshi, A.Y.; Joshi, A.Y. A systematic review on powder mixed electrical discharge machining. *Heliyon* **2019**, *5*, e02963. [CrossRef]
40. Zain, Z.M.; Ndaliman, M.B.; Khan, A.A.; Ali, M. Improving micro-hardness of stainless steel through powder-mixed electrical discharge machining. *Proc. Inst. Mech. Eng. Part C J. Mech. Eng. Sci.* **2014**, *228*, 3374–3380. [CrossRef]
41. Zain, Z.M.; Ndaliman, M.B.; Khan, A.A.; Ali, M.Y. Electro-Discharge Machining of SUS 304 Stainless Steel with TaC Powder-Mixed Dielectric. *Adv. Mater. Res.* **2012**, *576*, 72–75. [CrossRef]
42. Wang, J.L.; Yang, H.; Li, M. Study on Discharge Parameters of Surface Strengthening with Powder Mixed Near Dry EDM for H13 Steel. *Appl. Mech. Mater.* **2014**, *602–605*, 757–760. [CrossRef]
43. Wu, K.L.; Yan, B.H.; Huang, F.Y.; Chen, S.C. Improvement of surface finish on SKD steel using electro-discharge machining with aluminum and surfactant added dielectric. *Int. J. Mach. Tools Manuf.* **2005**, *45*, 1195–1201. [CrossRef]
44. Yan, B.H.; Tsai, H.C.; Huang, F.Y. The effect in EDM of a dielectric of a urea solution in water on modifying the surface of titanium. *Int. J. Mach. Tools Manuf.* **2005**, *45*, 194–200. [CrossRef]
45. Kumar, S.; Singh, R. Investigating surface properties of OHNS die steel after electrical discharge machining with manganese powder mixed in the dielectric. *Int. J. Adv. Manuf. Technol.* **2010**, *50*, 625–633. [CrossRef]
46. Kansal, H.; Singh, S.; Kumar, P. Effect of Silicon Powder Mixed EDM on Machining Rate of AISI D2 Die Steel. *J. Manuf. Process.* **2007**, *9*, 13–22. [CrossRef]
47. Tan, P.C.; Yeo, S.H. Investigation of recast layers generated by a powder-mixed dielectric micro electrical discharge machining process. *Proc. Inst. Mech. Eng. Part B J. Eng. Manuf.* **2011**, *225*, 1051–1062. [CrossRef]
48. Assarzadeh, S.; Ghoreishi, M. A dual response surface-desirability approach to process modeling and optimization of Al_2O_3 powder-mixed electrical discharge machining (PMEDM) parameters. *Int. J. Adv. Manuf. Technol.* **2012**, *64*, 1459–1477. [CrossRef]
49. Singh, B.; Kumar, J.; Kumar, S. Experimental Investigation on Surface Characteristics in Powder-Mixed Electrodischarge Machining of AA6061/10%SiC Composite. *Mater. Manuf. Process.* **2014**, *29*, 287–297. [CrossRef]
50. Wu, K.L.; Yan, B.H.; Lee, J.-W.; Ding, C.G. Study on the characteristics of electrical discharge machining using dielectric with surfactant. *J. Mater. Process. Technol.* **2009**, *209*, 3783–3789. [CrossRef]
51. Chen, Y.-F.; Lin, Y.-C. Surface modifications of Al–Zn–Mg alloy using combined EDM with ultrasonic machining and addition of TiC particles into the dielectric. *J. Mater. Process. Technol.* **2009**, *209*, 4343–4350. [CrossRef]
52. Lin, Y.C.; Yan, B.H.; Huang, F.Y. Surface modification of Al–Zn–Mg aluminum alloy using the combined process of EDM with USM. *J. Mater. Process. Technol.* **2001**, *115*, 359–366. [CrossRef]

53. Furutania, K.; Saneto, A.; Takezawa, H.; Mohri, N.; Miyake, H. Accretion of titanium carbide by electrical discharge machining with powder suspended in working fluid. *Precis. Eng.* **2001**, *25*, 138–144. [CrossRef]
54. Wang, Y.G.; Zhao, F.L.; Liu, Y. Behaviors of Suspended Powder in Powder Mixed EDM. *Key Eng. Mater.* **2008**, *375–376*, 36–41. [CrossRef]
55. Czelusniak, T.; Higa, C.F.; Torres, R.D.; Laurindo, C.A.H.; Júnior, J.M.F.D.P.; Lohrengel, A.; Amorim, F.L. Materials used for sinking EDM electrodes: A review. *J. Braz. Soc. Mech. Sci. Eng.* **2018**, *41*, 14. [CrossRef]
56. Peta, K.; Mendak, M.; Bartkowiak, T. Discharge Energy as a Key Contributing Factor Determining Microgeometry of Aluminum Samples Created by Electrical Discharge Machining. *Crystals* **2021**, *11*, 1371. [CrossRef]
57. Liu, Q.; Zhang, Q.; Zhang, M.; Yang, F. Study on the Discharge Characteristics of Single-Pulse Discharge in Micro-EDM. *Micromachines* **2020**, *11*, 55. [CrossRef] [PubMed]
58. Amin, A.; Rani, A.M.A.; Aliyu, A.A.A.; Razak, M.A.A.; Hastuty, S.; Bryant, M.G. Powder mixed-EDM for potential biomedical applications: A critical review. *Mater. Manuf. Process.* **2020**, *35*, 1789–1811. [CrossRef]
59. Bisaria, H.; Patra, B.B.; Mohanty, S. Surface modification during hydroxyapatite powder mixed electric discharge machining of metallic biomaterials: A review. *Surf. Eng.* **2022**, 1–27. [CrossRef]
60. Beig, B.; Liaqat, U.; Niazi, M.; Douna, I.; Zahoor, M.; Niazi, M. Current Challenges and Innovative Developments in Hydroxyapatite-Based Coatings on Metallic Materials for Bone Implantation: A Review. *Coatings* **2020**, *10*, 1249. [CrossRef]
61. Tyagi, R.; Mandal, A.; Das, A.K.; Tripathi, A.; Prakash, C.; Campilho, R.; Saxena, K.K. Electrical Discharge Coating a Potential Surface Engineering Technique: A State of the Art. *Processes* **2022**, *10*, 1971. [CrossRef]
62. Bains, P.S.; Bahraminasab, M.; Sidhu, S.S.; Singh, G. On the machinability and properties of Ti–6Al–4V biomaterial with n-HAp powder–mixed ED machining. *Proc. Inst. Mech. Eng. Part H J. Eng. Med.* **2019**, *234*, 232–242. [CrossRef]
63. Bui, V.D.; Mwangi, J.W.; Meinshausen, A.-K.; Mueller, A.J.; Bertrand, J.; Schubert, A. Antibacterial coating of Ti-6Al-4V surfaces using silver nano-powder mixed electrical discharge machining. *Surf. Coat. Technol.* **2019**, *383*, 125254. [CrossRef]
64. Prakash, C.; Kansal, H.K.; Pabla, B.S.; Puri, S. Experimental investigations in powder mixed electric discharge machining of Ti–35Nb–7Ta–5Zrβ-titanium alloy. *Mater. Manuf. Process.* **2016**, *32*, 274–285. [CrossRef]
65. Shabgard, M.; Khosrozadeh, B. Investigation of carbon nanotube added dielectric on the surface characteristics and machining performance of Ti–6Al–4V alloy in EDM process. *J. Manuf. Process.* **2017**, *25*, 212–219. [CrossRef]
66. Hourmand, M.; Sarhan, A.A.D.; Farahany, S.; Sayuti, M. Microstructure characterization and maximization of the material removal rate in nano-powder mixed EDM of Al-Mg_2Si metal matrix composite—ANFIS and RSM approaches. *Int. J. Adv. Manuf. Technol.* **2018**, *101*, 2723–2737. [CrossRef]
67. Santosh, S.; Ibrahim, S.J.S.; Saravanamuthukumar, P.; Rajkumar, K.; Harikrishna, K.L. Nano Graphite Powder Assisted Electric Discharge Machining Characteristics of ZM21 Magnesium Alloy. *Appl. Mech. Mater.* **2015**, *787*, 406–410. [CrossRef]
68. Lamichhane, Y.; Singh, G.; Bhui, A.S.; Mukhiya, P.; Kumar, P.; Thapa, B. Surface modification of 316L SS with HAp nano-particles using PMEDM for enhanced Biocompatibility. *Mater. Today Proc.* **2019**, *15*, 336–343. [CrossRef]
69. Banh, T.-L.; Nguyen, H.-P.; Ngo, C.; Nguyen, D.-T. Characteristics optimization of powder mixed electric discharge machining using titanium powder for die steel materials. *Proc. Inst. Mech. Eng. Part E J. Process. Mech. Eng.* **2017**, *232*, 281–298. [CrossRef]
70. Prakash, C.; Uddin, M. Surface modification of β-phase Ti implant by hydroaxyapatite mixed electric discharge machining to enhance the corrosion resistance and in-vitro bioactivity. *Surf. Coat. Technol.* **2017**, *326*, 134–145. [CrossRef]
71. Abdul-Rani, A.; Nanimina, A.; Ginta, T.; Razak, M. Machined Surface Quality in Nano Aluminum Mixed Electrical Discharge Machining. *Procedia Manuf.* **2017**, *7*, 510–517. [CrossRef]
72. Prakash, C.; Kansal, H.; Pabla, B.; Puri, S. Potential of Silicon Powder-Mixed Electro Spark Alloying for Surface Modification of β-Phase Titanium Alloy for Orthopedic Applications. *Mater. Today Proc.* **2017**, *4*, 10080–10083. [CrossRef]
73. Farooq, M.U.; Mughal, M.P.; Ahmed, N.; Mufti, N.A.; Al-Ahmari, A.M.; He, Y. On the Investigation of Surface Integrity of Ti6Al4V ELI Using Si-Mixed Electric Discharge Machining. *Materials* **2020**, *13*, 1549. [CrossRef] [PubMed]
74. Devgan, S.; Sidhu, S.S. Surface modification of β-type titanium with multi-walled CNTs/μ-HAp powder mixed Electro Discharge Treatment process. *Mater. Chem. Phys.* **2020**, *239*, 122005. [CrossRef]
75. Razak, M.A.; Rani, A.M.A.; Saad, N.M.; Littlefair, G.; Aliyu, A.A. Controlling corrosion rate of Magnesium alloy using powder mixed electrical discharge machining. *IOP Conf. Ser. Mater. Sci. Eng.* **2018**, *344*, 012010. [CrossRef]
76. Rout, P.K.; Surekha, B.; Jena, P.C.; Arko, G.N. Experimental studies on aluminum powder mixed electro discharge machining of ultra-light weight Mg-6Al alloy. *Mater. Today Proc.* **2020**, *26*, 2379–2387. [CrossRef]
77. Mohan, D.G.; Gopi, S.; Rajasekar, V.; Krishnan, K.; Mohan, D.G.; Gopi, S.; Selvarajan, L.; Rajavel, R.; Prakash, B.; Mohan, D.G.; et al. Accepted Mus pt. *Mater. Today Proc.* **2019**, *27*. [CrossRef]
78. Singh, G.; Lamichhane, Y.; Bhui, A.S.; Sidhu, S.S.; Bains, P.S.; Mukhiya, P. Surface morphology and microhardness behavior of 316l in hap-pmedm. *Facta Univ. Ser. Mech. Eng.* **2019**, *17*, 445–454. [CrossRef]
79. Bhaumik, M.; Maity, K. Effect of machining parameter on the surface roughness of AISI 304 in silicon carbide powder mixed EDM. *Decis. Sci. Lett.* **2017**, 261–268. [CrossRef]
80. Kansal, H.; Singh, S.; Kumar, P. Application of Taguchi method for optimisation of powder mixed electrical discharge machining. *Int. J. Manuf. Technol. Manag.* **2005**, *7*, 329. [CrossRef]
81. Ishfaq, K.; Asad, M.; Anwar, S.; Pruncu, C.I.; Saleh, M.; Ahmad, S. A Comprehensive Analysis of the Effect of Graphene-Based Dielectric for Sustainable Electric Discharge Machining of Ti-6Al-4V. *Materials* **2020**, *14*, 23. [CrossRef]

82. Garg, R.K.; Ojha, K. Parametric Optimization of PMEDM Process with Chromium Powder Suspended Dielectric for Triangular Electrodes. *Adv. Mater. Res.* **2013**, *816–817*, 23–27. [CrossRef]
83. Aliyu, A.A.; Abdul-Rani, A.M.; Rubaiee, S.; Danish, M.; Bryant, M.; Hastuty, S.; Razak, M.A.; Ali, S. Electro-Discharge Machining of $Zr_{67}Cu_{11}Ni_{10}Ti_9Be_3$: An Investigation on Hydroxyapatite Deposition and Surface Roughness. *Processes* **2020**, *8*, 635. [CrossRef]
84. Bhattacharya, A.; Batish, A.; Singh, G. Optimization of powder mixed electric discharge machining using dummy treated experimental design with analytic hierarchy process. *Proc. Inst. Mech. Eng. Part B J. Eng. Manuf.* **2011**, *226*, 103–116. [CrossRef]
85. Talla, G.; Gangopadhyay, S.; Biswas, C.K. Multi Response Optimization of Powder Mixed Electric Discharge Machining of Aluminum/Alumina Metal Matrix Composite Using Grey Relation Analysis. *Procedia Mater. Sci.* **2014**, *5*, 1633–1639. [CrossRef]
86. Amin, A.; Abdul-Rani, A.M.; Ahmed, R.; Rao, T.V.V.L.N. Multiple-objective optimization of hydroxyapatite-added EDM technique for processing of 316L-steel. *Mater. Manuf. Process.* **2021**, 1–12. [CrossRef]
87. Amin, A.; Abdul-Rani, A.M.; Ahmed, R.; Shahid, M.U.; Zohura, F.T.; Rani, M.D.B.A. Multi-objective optimization of process variables for MWCNT-added electro-discharge machining of 316L steel. *Int. J. Adv. Manuf. Technol.* **2021**, *115*, 179–198. [CrossRef]
88. Padhee, S.; Nayak, N.; Panda, S.K.; Dhal, P.R.; Mahapatra, S.S. Multi-objective parametric optimization of powder mixed electro-discharge machining using response surface methodology and non-dominated sorting genetic algorithm. *Sadhana* **2012**, *37*, 223–240. [CrossRef]
89. Tripathy, S.; Tripathy, D. Multi-attribute optimization of machining process parameters in powder mixed electro-discharge machining using TOPSIS and grey relational analysis. *Eng. Sci. Technol. Int. J.* **2016**, *19*, 62–70. [CrossRef]
90. Prakash, C.; Kansal, H.K.; Pabla, B.S.; Puri, S. Multi-objective optimization of powder mixed electric discharge machining parameters for fabrication of biocompatible layer on β-Ti alloy using NSGA-II coupled with Taguchi based response surface methodology. *J. Mech. Sci. Technol.* **2016**, *30*, 4195–4204. [CrossRef]
91. Kansal, H.; Singh, S.; Kumar, P. Numerical simulation of powder mixed electric discharge machining (PMEDM) using finite element method. *Math. Comput. Model.* **2008**, *47*, 1217–1237. [CrossRef]
92. Bhattacharya, A.; Batish, A.; Singh, K. Fe simulation and experimental validation of powder mixed edm process for estimating the temperature distribution and volume removed in single crater. *Int. J. Model. Simul. Sci. Comput.* **2012**, *3*, 1250006. [CrossRef]
93. Bhattacharya, A.; Batish, A. Predictor Equations for Estimating Crater Dimensions in PMEDM Process Using Fem Simulation and Experimental Validation. *Mater. Sci. Forum* **2013**, *751*, 45–60. [CrossRef]
94. Tan, P.C.; Yeo, S.H. Simulation of Surface Integrity for Nanopowder-Mixed Dielectric in Micro Electrical Discharge Machining. *Met. Mater. Trans. B* **2013**, *44*, 711–721. [CrossRef]
95. Wandra, R. Finite element analysis of powder mixed electric discharge machining on Ti-alloy. *Mater. Today Proc.* **2021**, *50*, 593–599. [CrossRef]
96. Tümer, D.; Güngörürler, M.; Havıtçıoğlu, H.; Arman, Y. Investigation of effective coating of the Ti–6Al–4V alloy and 316L stainless steel with graphene or carbon nanotubes with finite element methods. *J. Mater. Res. Technol.* **2020**, *9*, 15880–15893. [CrossRef]
97. Jampana, V.N.R.; Rao, P.S.V.R.; Sampathkumar, A. Experimental and Thermal Investigation on Powder Mixed EDM Using FEM and Artificial Neural Networks. *Adv. Mater. Sci. Eng.* **2021**, *2021*, 8138294. [CrossRef]
98. Bhattacharya, A.; Batish, A.; Singh, G.; Singla, V.K. Optimal parameter settings for rough and finish machining of die steels in powder-mixed EDM. *Int. J. Adv. Manuf. Technol.* **2011**, *61*, 537–548. [CrossRef]
99. Mohri, N.; Saito, N.; Higashi, M.; Kinoshita, N. A New Process of Finish Machining on Free Surface by EDM Methods. *CIRP Ann.* **1991**, *40*, 207–210. [CrossRef]
100. Peças, P.; Henriques, E. Influence of silicon powder-mixed dielectric on conventional electrical discharge machining. *Int. J. Mach. Tools Manuf.* **2003**, *43*, 1465–1471. [CrossRef]
101. Kibria, G.; Bhattacharyya, B. Investigation into micro-hole geometrical accuracy during micro-EDM of Ti-6Al-4V employing different dielectrics. *Int. J. Mach. Mach. Mater.* **2011**, *10*, 310. [CrossRef]
102. Kibria, G.; Bhattacharyya, B.; Chinesta, F.; Chastel, Y.; El Mansori, M. Analysis on Geometrical Accuracy of Microhole during Micro-EDM of Ti-6Al-4V using Different Dielectrics. *AIP Conf. Proc.* **2011**, *1315*, 155. [CrossRef]
103. Chen, S.-L.; Lin, M.-H.; Huang, G.-X.; Wang, C.-C. Research of the recast layer on implant surface modified by micro-current electrical discharge machining using deionized water mixed with titanium powder as dielectric solvent. *Appl. Surf. Sci.* **2014**, *311*, 47–53. [CrossRef]
104. Bhattacharya, A.; Batish, A.; Kumar, N. Surface characterization and material migration during surface modification of die steels with silicon, graphite and tungsten powder in EDM process. *J. Mech. Sci. Technol.* **2013**, *27*, 133–140. [CrossRef]
105. Kucukturk, G.; Cogun, C. A new method for machining of electrically nonconductive workpieces using electric discharge machining technique. *Mach. Sci. Technol.* **2010**, *14*, 189–207. [CrossRef]
106. Srivastava, S.; Vishnoi, M.; Gangadhar, M.T.; Kukshal, V. An insight on Powder Mixed Electric Discharge Machining: A state of the art review. *Proc. Inst. Mech. Eng. Part B J. Eng. Manuf.* **2022**. [CrossRef]
107. Al-Amin, M.; Abdul-Rani, A.M.; Danish, M.; Rubaiee, S.; Mahfouz, A.B.; Thompson, H.M.; Ali, S.; Unune, D.R.; Sulaiman, M.H. Investigation of Coatings, Corrosion and Wear Characteristics of Machined Biomaterials through Hydroxyapatite Mixed-EDM Process: A Review. *Materials* **2021**, *14*, 3597. [CrossRef]
108. Philip, J.T.; Mathew, J.; Kuriachen, B. Transition from EDM to PMEDM—Impact of suspended particulates in the dielectric on Ti_6Al_4V and other distinct material surfaces: A review. *J. Manuf. Process.* **2021**, *64*, 1105–1142. [CrossRef]

109. Al Amin, M.; Rani, A.M.A.; Aliyu, A.A.A.; Bryant, M.G.; Danish, M.; Ahmad, A. Bio-ceramic coatings adhesion and roughness of biomaterials through PM-EDM: A comprehensive review. *Mater. Manuf. Process.* **2020**, *35*, 1157–1180. [CrossRef]
110. Abdudeen, A.; Abu Qudeiri, J.E.; Kareem, A.; Ahammed, T.; Ziout, A. Recent Advances and Perceptive Insights into Powder-Mixed Dielectric Fluid of EDM. *Micromachines* **2020**, *11*, 754. [CrossRef]
111. Aliyu, A.A.; Abdul-Rani, A.M.; Ginta, T.L.; Prakash, C.; Axinte, E.; Razak, M.A.; Ali, S. A Review of Additive Mixed-Electric Discharge Machining: Current Status and Future Perspectives for Surface Modification of Biomedical Implants. *Adv. Mater. Sci. Eng.* **2017**, *2017*, 8723239. [CrossRef]
112. Marashi, H.; Jafarlou, D.M.; Sarhan, A.A.; Hamdi, M. State of the art in powder mixed dielectric for EDM applications. *Precis. Eng.* **2016**, *46*, 11–33. [CrossRef]
113. Bajaj, R.; Tiwari, A.K.; Dixit, A.R. Current Trends in Electric Discharge Machining Using Micro and Nano Powder Materials—A Review. *Mater. Today Proc.* **2015**, *2*, 3302–3307. [CrossRef]

machines

Article

Enhancing Dimensional Accuracy in Budget-Friendly 3D Printing through Solid Model Geometry Tuning and Its Use in Rapid Casting

Barun Haldar

Mechanical Engineering Department, College of Engineering, Imam Mohammad Ibn Saud Islamic University (IMSIU), Riyadh 11432, Saudi Arabia; bhaldar@imamu.edu.sa

Abstract: Achieving precise dimensional accuracy and improving surface quality are the primary research and development objectives in the engineering and industrial applications of 3D printing (3DP) technologies. This experimental study investigates the pivotal role of solid model geometry tuning in enhancing the dimensional accuracy of affordable 3D printing technologies, with a specific focus on economical engineering applications. This experiment utilises low-cost Material Extrusion/Fused Filament Fabrication (FFF) and Stereolithography (SLA)/Digital Light Processing (DLP) 3D-printed patterns for the meticulous measurement of errors in the X, Y, and Z directions. These errors are then used to refine subsequent solid models, resulting in a marked improvement in dimensional accuracy (i.e., 0.15%, 0.33%, and 2.16% in the X, Y, and Z directions, respectively) in the final DLP 3D-printed parts. The study also derives and experimentally validates a novel and simple mathematical model for tuning the solid model based on the calculated linear directional errors (e_i, e_j, and e_k). The developed mathematical model offers a versatile approach for achieving superior dimensional accuracy in other 3D printing processes. Medium-sized (4 to 10 cm) wax-made DLP- and PLA-made patterns are used to test the ceramic mould-building capacity for rapid casting (RC), where the FFF-based 3D-printed (hollow inside) pattern favours successful RC. This work comprehensively addresses the critical challenges encountered in low-cost DLP and FFF processes and their scopes in engineering applications. It provides novel suggestions and answers to improve the effectiveness, quality, and accuracy of the FFF 3D printing process for future applications in RC.

Keywords: low-cost 3D printing; additive manufacturing; dimensional accuracy enhancing; rapid casting; Industry 4.0

1. Introduction

1.1. The 3D Printing Technology and Its Industrial Significance

Basic 3D printing (3DP), a layer-by-layer generative manufacturing process, also known as rapid prototyping (RP), began in the 1980s with the innovation of 3D Systems Inc. CA [1]. Due to its accessibility, 3DP was initially utilised for prototyping by architects and designers. Its usage has now extended to schools, homes, offices, libraries, and labs [2], primarily due to its affordability. The technological and material evaluation tends to the direct manufacturing (DM) [2] of lightweight, strong, and safe products just-in-time with cost benefits. To meet the demands of the ongoing industrial revolution, additive manufacturing (AM)/3DP technologies are booming areas of research and enabling the production of exotic industrial products and even artificial human organs directly from the designed 3D model. The ability to manufacture integrated and complex geometries (which are impossible to produce using conventional manufacturing processes) [3], with high precision, material savings, and design flexibility [4], represents the key features of this cutting-edge technology [5], making it a fundamental component of Industry 4.0 [6]. It offers the potential for mass customisation and personalisation, thereby providing an opportunity to eliminate high custom-tailored production costs [4]. Moreover, it enhances response time, shortens supply chains, cuts storage and delivery costs, and reduces lead times for crucial replacement parts in on-demand manufacturing [3] according to historical global survey statistics and forecasts on yearly spending for final 3DP component manufacturing provided by Hubs, Wohler's Associates, and SmarTech [7–11] are depicted in Figure 1.

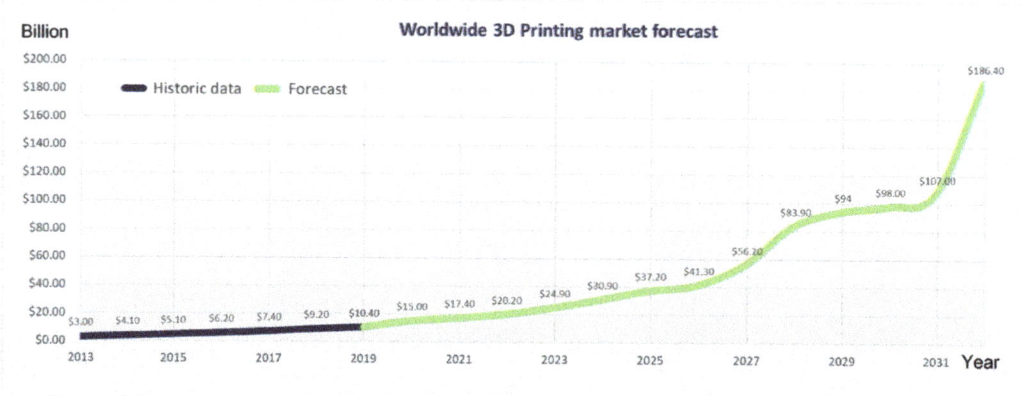

Figure 1. Projection of additive manufacturing (AM) worldwide market values [7–11].

AM is rapidly expanding in today's industrial sectors, and the demands of AM have increased almost fourfold in the last five years. Out of the total parts manufactured by 3DP worldwide in 2020 (as plotted in Figure 2), 47% were prototypes, 21% were jigs and fixtures, 29% were aesthetic and functional parts, and 3% were others (data source: Additive manufacturing trend report 2021, Hubs) [8].

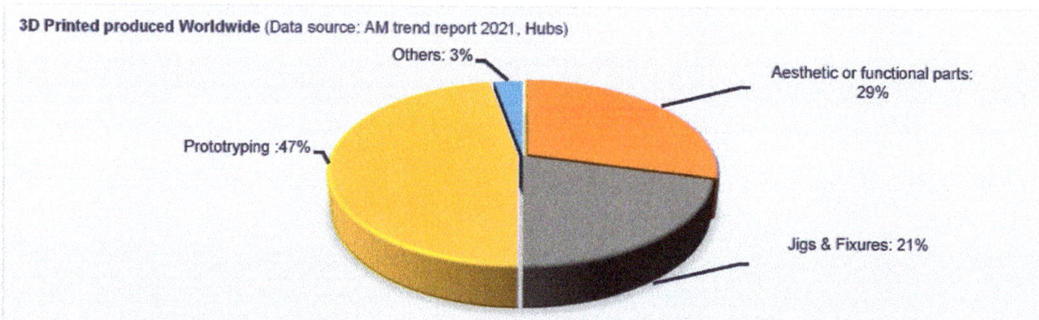

Figure 2. Worldwide applications of AM products [8].

The various AM methods' in-built rate vs. machine cost are presented in Figure 3.

Among the various 3DP technologies available, four main additive manufacturing methods stand out: fused deposition modelling (FDM)/material extrusion, binder jetting, stereolithography(SLA), and selective laser sintering (SLS) [4]. In direct metal 3DP/AM processes, the price of the laser and the cost of the raw ingredients (metal powders) make the abovementioned methods expensive to invest in. Industrial-grade metal 3D printers can range in price from several hundred thousand to a few million USD [12], while FDM or SLA printers typically range from a few hundred to a few thousand USD. Figure 4 displays the global share of metal-based AM technologies sold in 2020.

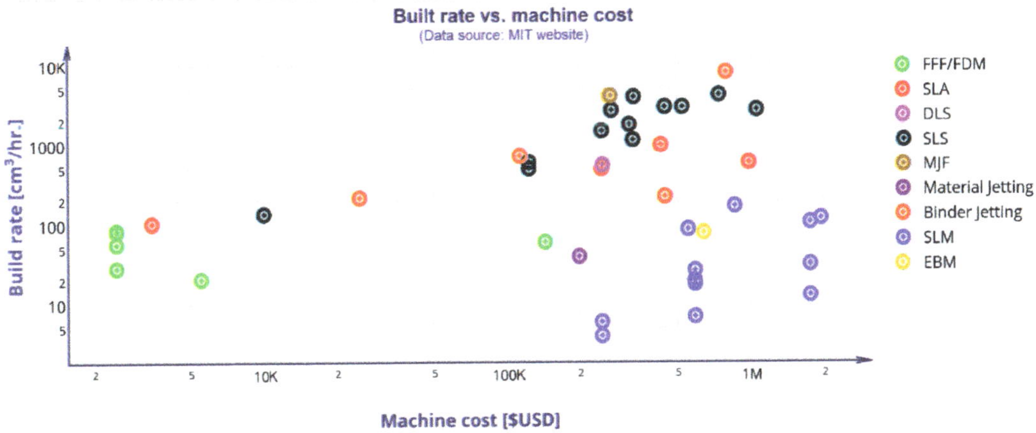

Figure 3. Comparing AM processes in-built rate vs. machine cost [12].

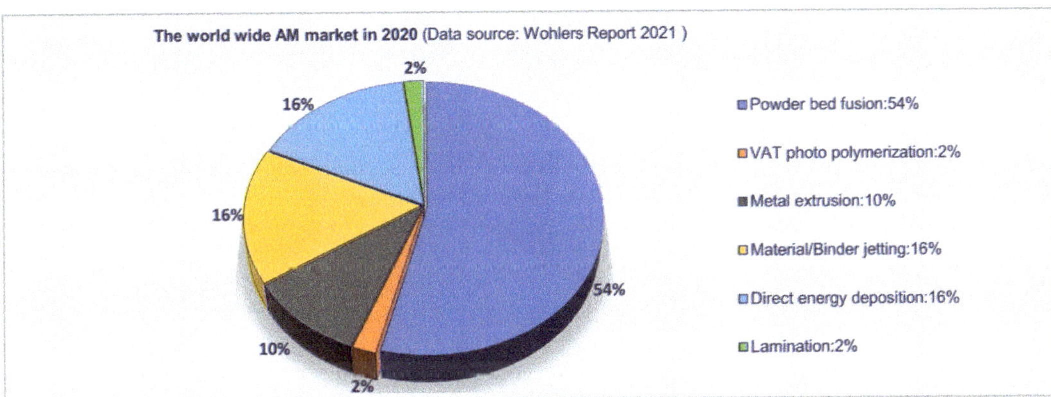

Figure 4. The worldwide metal additive manufacturing market in 2020 [3].

On the verge of the growing popularity of low-cost 3DP technologies, FDM/FFF and SLA are the two leading technologies in consideration. In addition, these two technologies are promising for utilisation in the pattern making of RC to produce metallic industrial products in an indirect route. The comparative characteristics of direct metal 3DP vs. polymer and photo-polymer-based low cost (FFF/SLA) are tabulated in Table 1.

Table 1. Some comparative critical characteristics of direct 3DP and FFF or SLA cum ceramic casting.

Characteristics	Metallic Product Manufacturing Through:	
	FDM/FFF or SLA Patterns	**Direct 3DP**
Advantages:	▪ Low cost (polymer) of products, prototypes, and patterns 3DP ▪ Low- to medium-complexity metal products can be manufactured by casting using 3D-printed patterns ▪ Customised product manufacturing ▪ Flexible in metal, alloys, super alloys, product making, etc.	▪ Direct metal product 3DP ▪ High-complexity product-building capacity ▪ Customised product manufacturing ▪ Wide range of metals and alloys using scopes ▪ Scope for product weight reduction (by producing internal mesh structure) for efficient exotic applications like aerospace and medical implants…

Table 1. *Cont.*

Characteristics	Metallic Product Manufacturing Through:	
	FDM/FFF or SLA Patterns	Direct 3DP
Limitations:	▪ Costly setup, peripherals, and raw materials ▪ High running and maintenance cost ▪ Limited manufacturer of metal 3DP machines	▪ Need skilled staffing ▪ Higher production time (due to 3D printing and ceramic/investment casting) ▪ A metallic internal mesh structure is not possible
Research scope:	▪ Dimensional accuracy and surface topography [13] ▪ Microstructure structure control and defects ▪ Artificial intelligence (AI) integration for process control and quality improvement ▪ Atomic self-assembly [14] or area-selective atomic layer deposition [15] for faster and quality production ▪ Use of difficult-to-machine exotic materials like MMCs, super alloys and nano-materials ▪ Quality improvement and cost minimisation [1]	▪ Dimensional accuracy and surface topography [1] ▪ Printing speed and size [1] improvement ▪ Widening range of 3DP materials ▪ Use of difficult-to-machine exotic materials like MMCs, super alloys, etc. ▪ Quality improvement and cost minimisation [1]

Note: Material extrusion/FFF technology is also under development for direct metal product manufacturing [16].

Advanced metal 3DP (Selective Laser Melting (SLM)) technologies, namely Direct Metal Laser Sintering (DMLS), Selective Laser Sintering (SLS), Direct Metal Printing (DMP), Laser Powder Bed Fusion (LPBF), and Selective Laser Melting (SLM) [17], use expensive energy sources like laser, electron beam, and costly metal powder for obtaining an exotic metal product through 3DP. This technology allows for free-form design with higher complexity and enables the manufacture of items that no other manufacturing process can produce. Shapes with lower to moderate complexity can be cost-effective through indirect or alternative methods, such as RC. Continuous research and development in FMD/FFF and SLA 3DP technologies are necessary for such benefits to be obtained in industrial contexts. The global demand trends of using polymer and photopolymer in AM were reported (2022) by Wohlers Association [18] and are plotted here in Figure 5.

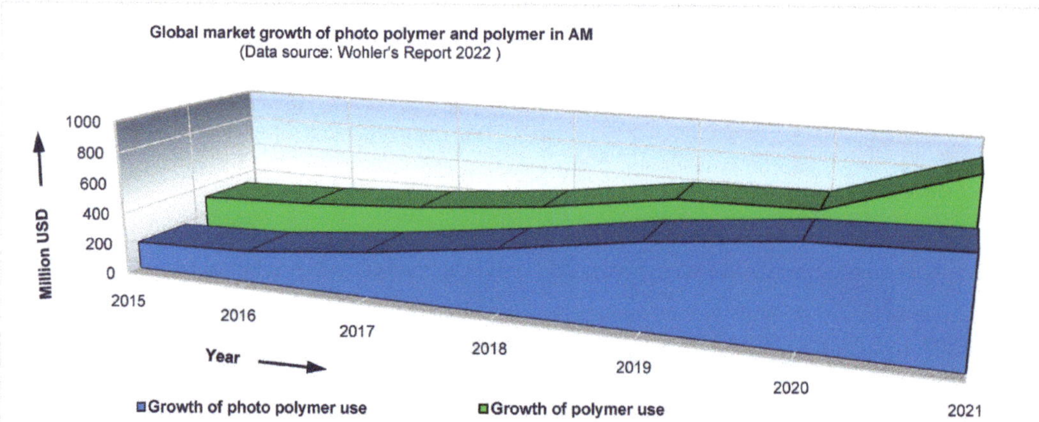

Figure 5. Growth of the polymer and photopolymer market worldwide [18].

With the improved dimensional accuracy of low-cost FFF and SLA 3D printing processes, these technologies could gain more popularity in the investment casting of exotic materials and alloys, potentially facilitating cellular manufacturing and encouraging entrepreneurship. To assist in selecting an affordable Fused Filament Fabrication (FFF) or Stereolithography (SLA) process, the comparative characteristics of these processes are provided in Table 2.

Table 2. The characteristics of FFF and SLA (DLP) 3DP.

Characteristics	FDM/FFF process	DLP 3DP
Cost:	▪ Mostly available low-cost printer	▪ Available low-cost printer
Consumable:	▪ Filament	▪ Vat resin, fluorinated ethylene propylene (FEP) film
Running skills:	▪ User friendly	▪ The skilled operator is essential in changing FEP film and VAT resin.
Health issue:	▪ Safe in handling ▪ No smell	▪ VAT resin is toxic and detrimental to respiratory and cardiovascular [19] organs
Maintenance:	▪ Easier	▪ Relatively difficult
Running cost:	▪ Low	▪ Relatively high
Accuracy:	▪ Poor	▪ Poor
Surface quality:	▪ Poor due to wavy surface and stair-casing effects [Figure 6a]	▪ Sooth horizontal and vertical surfaces but sharp stair-casing results in slanting characters [Figure 6b].
Complexity:	▪ Highly complex geometry builds possible	▪ Suitable for low- to medium complexity geometry
Material savings:	▪ Inside mesh/honeycomb structure [Figure 6a] can save material, and it is favourable to accommodate thermal stress in RC ceramic shell burning	▪ Inert [Figure 6b] and material saving is not possible

The layer-by-layer material build mechanisms of FFF and DLP 3DP processes are shown in Figure 6a,b, respectively.

The mesh/honeycomb building capacity of FFF 3DP under the surfaces of the workpiece is adequate for material savings, while using it as a pattern in RC, it can also easily accommodate thermal expansion due to burning of the ceramic mould for cavity making. On the other hand, the SLA process is well established for minimal non-engineering product RC, like, making of jewellery, dentistry [20–23], items where dimensional tolerance is not mandatory.

The improvement in the dimensional accuracy of such low-cost FFF and SLA processes and experimental investigations to use such patterns in RC may beget effective outcomes for the successful engineering implementation of such technology. This may significantly reduce the initial investment cost for manufacturing metallic products in the rapid casting (RC) route, which is favourable for fostering entrepreneurship development and industrial applications.

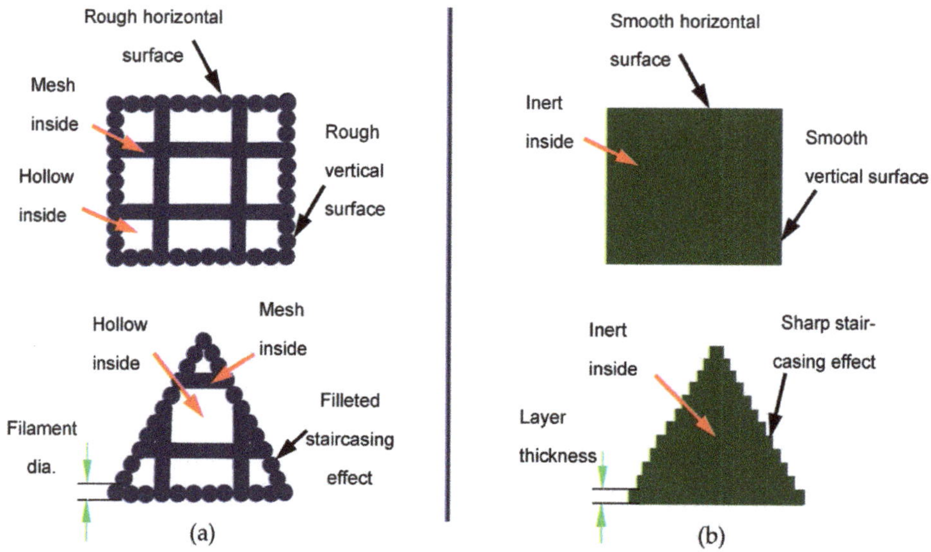

Figure 6. The layer-by-layer material build mechanisms of (**a**) FFF and (**b**) DLP 3D printing processes.

1.2. Integration of 3DP in Rapid Casting in Industrial Contexts

The World Casting Census 2019, published by the American Foundry Society in 2021, reported the existence of more than 45,377 [24] well-functioning metal casting plants worldwide. Approximately 10% of these plants produce high-quality steel products through the ceramic shell and/or investment casting processes. These processes are near-net-shaped manufacturing methods that enable the production of complex metal shapes with a high dimensional accuracy and excellent surface finish [25], often without the need for post-processing. The main advantages of conventional IC are [26]:

- Shape complexity: almost any degree of external complexity and a wide range of internal complexity can be achieved;
- Freedom of alloy selection: super alloys, MMC, advanced materials, etc.;
- Close dimensional tolerances: consistent and repetitive close tolerances [27] and accuracy grade between CT4 and CT8 [28];
- The availability of prototype and temporary tooling;
- Reliability: demanding industries, including gas turbine engines, petroleum, chemicals, military, and medicine, have long relied on investment casting [27];
- Wide range of applications: few grams to more than 300 kg.

The ceramic shell mould can withstand high pouring temperatures (1500 °C to 1600 °C) [29], which makes the process compatible with casting high-temperature melting point metals, such as Y_2O_3, $CaZrO_3$, etc. Ceramic shells [30,31] enable the casting of metals like titanium (with a melting point above 1600 °C) and superalloys.

The hybridisation of investment casting using a low-cost 3DP (FFF, SLA) pattern can overcome the use of the high investment cost (varies from 100K to 2M USD) [12] of metal 3DP.

Additionally, the hybridisation of cost-effective 3D printing (FFF, SLA) with investment casting has the potential to eliminate mutual limitations and yield combined advantages from both processes. The key characteristics of low-cost 3D printing (FFF, SLA), investment casting (IC), and RC (hybrid) are summarised in Table 3.

Table 3. The characteristics of low-cost 3D printing (FFF, SLA), investment casting (IC), and RC [1].

Characteristics	Low-Cost (FFF, SLA) 3DP (AM) Technologies	Investment Casting	Rapid Investment Casting/Rapid Casting
Cost:	Low	Metallic die price is affordable in mass production only.	Profitable for job production to batch production, reverse engineering, just-in-time, and cellular manufacturing
Material:	Low melting points like PLA, PVA, WAX, etc.	Metals, alloys, super alloys	Metals, alloys, super alloys
Production time:	1 to 7 days for a batch of product	Significant time consumed initially for making metallic dies	2-8 days for a batch of product
Geometry complexity:	FDM/FFF can make any complex internal geometry, and any externally connected internal geometry can be made using SLA 3DP	Any externally connected internal geometry can be made using SLA 3DP	Any externally linked internal geometry can be made using SLA 3DP
Just-in-time:	Possible	Not viable for small amounts and quick delivery	Possible
Cellular manufacturing:	Possible	NA	Possible
Reverse engineering:	Possible	NA	Possible

Finally, the augmentation of cost-effective metal products using RIC/RC (hybrid) will reduce the initial investment cost and production cost in some production ranges. Additionally, a wide variety of raw materials, including metals, alloys, superalloys, and metal matrix composites, can be used, thereby opening up further research opportunities. The expansion of metallic product manufacturing through rapid casting (RC) using Fused Filament Fabrication (FFF) or Stereolithography (SLA) patterns may present the following research and application opportunities, as depicted in the schematic gear diagram in Figure 7.

Aircraft engines, robot parts, airframes, aerospace, missiles, fuel systems, automotive, bicycles and motorcycles, materials handling equipment, ground support systems, agricultural equipment, textile equipment, baling and strapping equipment, dentistry and dental tools electrical equipment, cameras, electronics, radar, guns and small armaments, hand tools jewellery, machine tools, metalworking equipment, pneumatic and hydraulic systems, oil well drilling and auxiliary equipment, prosthetic appliances, high-pressure pumps, sports items, turbines, and wire processing equipment/parts [26] could all be effectively manufactured using the RC/RIC process.

The present state-of-the-art research status of hybrid 3DP and casting, i.e., RC, is presented in Table 4.

Table 4. The research status of RC (hybrid) and IC.

Researcher/s Details	Research Objectives	Materials	Results/Conclusion
(2021), F. Li et al. [32]	▪ SS impeller RC through 3DP of ceramic shell	▪ $Al_2O_3.2SiO_2$ (calcined kaolin) suspension and UV-curable resin binder, sintered at 1200 °C	▪ Shell work surface roughness 4.51~4.82 μm ▪ Post-treatment of the ceramic shell with a fine clay ceramic layer improved the surface quality of the parts

Table 4. *Cont.*

Researcher/s Details	Research Objectives	Materials	Results/Conclusion
(2020), M. Mukhtarkhanov et al. [33]	▪ RC through FEM and SLA 3DP	▪ Patterns: ABS, PLA, WAX	▪ Surface finish 83% enhanced due to the post-treatment of ABS and PLA patterns. ▪ Wax pattern tolerance limit up to 500 μm
(2019), UE Klotz et al. [31]	▪ IC of titanium alloys with	▪ Silica and calcium zirconate ($CaZrO_3$) shell, yttria face coat on alumina-based crucibles, $CaZrO_3$ crucibles	▪ $CaZrO_3$ showed stability as a shell and crucible in Ti casting
(2019), V. K. Tiwary et al. [34]	▪ FDM-based pattern for improving the surface quality of medical implants	▪ Pre- and post-processing of acrylonitrile butadiene styrene FDM pattern for IC of low alloy carbon steel	▪ Surface roughness decreased from 23.37 to 0.68 μm, dimensional divergence from 1 to 3%, and lead time reduced from one week to one day ▪ Polystyrene/wax/ PLA as FDM pattern materials could be used for comparative study
(2018), D. Wang et al. [35]	▪ Shorten the production cycle time of the IC impeller through simulation and optimisation of the gating system	▪ Wax patterns: high-impact polystyrene (HIPS) and photosensitive resin by SLS and SL processes. Mould shell: Silica and zircon clay	▪ Rapid IC reduces production lead time and costs and improves product quality ▪ The photosensitive resin is not suitable for IC
(2013), Y. Chen et al. [30]	▪ Thin-walled high-temperature Ti alloy IC	▪ Wax patterns made using an aluminium mould ▪ Y_2O_3, ZrO_2, and Al_2O_3 shell pre-heating at 300, 600, and 900 °C and filled, mesh pattern used for filling capacity evaluation	▪ High pre-heating temperature increases interfacial reaction, Y_2O_3 shell performs better at 300 °C pre-heating ▪ Using a 3D-printed pattern for making a Y_2O_3 shell and a comparative study could be performed in IC
(2012), M. Macků and M. Horáček [36]	▪ Dimensional changes study in FDM to IC of AlSi7Mg0.6	▪ FDM ABS pattern was used in the silicon die of the wax pattern making	▪ Dimensional variation in the final product is about 1 to 2% ▪ The use of direct wax pattern 3DP may further improve dimensional accuracy

Table 4. *Cont.*

Researcher/s Details	Research Objectives	Materials	Results/Conclusion
(2010), S. Wang et al. [37]	▪ Expandable plastic patterns used in IC	▪ The plastic pattern is burnt at 1120 °C to obtain a cavity in the ceramic shell	▪ Successful metal IC ▪ Plastic patterns burning may cause environment-related issues
(2011), M. Vaezi et al., 2011 [38]	▪ Reverse engineering using RC of turbine blades	▪ Wax patterns made by MultiJet Modelling (MJM) and CNC machined aluminium die	▪ MJM technology is cheaper and has a shorter lead time for job-shop production ▪ CNC-machined aluminium dies may be beneficial in mass production
(2007), E. Bassoli and A. Gatto [39]	▪ Dimensional accuracy in the cavity of light-alloys IC	▪ Starch patterns and ZCast process	▪ ZCast shows satisfactory results ▪ The ZCast process leaves parting line marks on the product surface

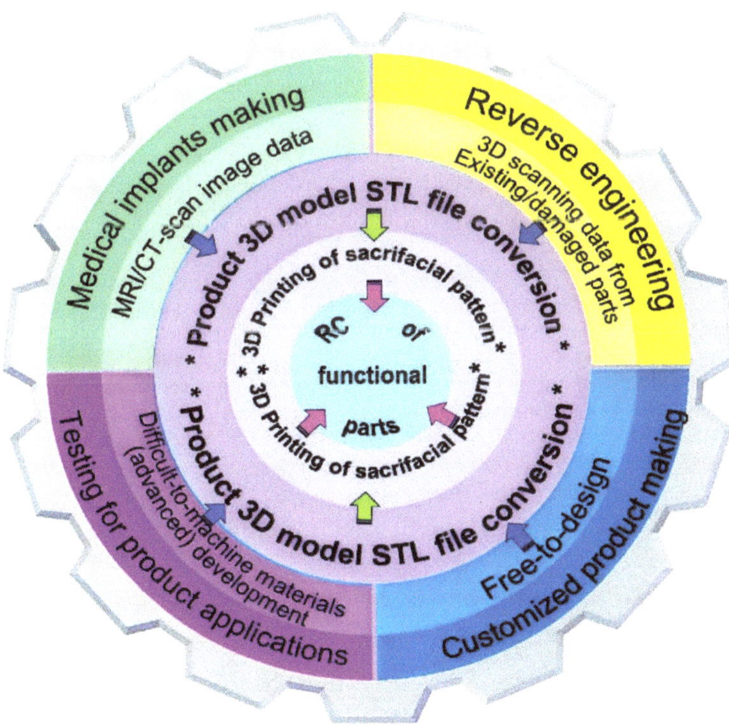

Figure 7. Research and the application scopes of RC/RIC through 3DP.

Searches and reviews indicate that there has been limited research conducted on the RC/hybridizing of 3DP and the precision casting process. The process has considerable potential to reduce the production cost from job shop to batch production levels. Unless the direct metal 3D printing process becomes cost-effective (which is challenging), the RC process may serve as an alternative, to some extent, for cost-effective production. Therefore, the use of affordable 3D printers has the potential to significantly reduce overall production costs if the dimensional accuracy of such 3D printing can be enhanced for engineering and industrial applications. The author of this work employed solid model geometry adjustments to enhance the dimensional accuracy of affordable FFF and SLA/DLP printing patterns. These adjustments were experimentally applied in ceramic casting to explore the potential for further development in this field. A schematic overview of the present work is given in Figure 8.

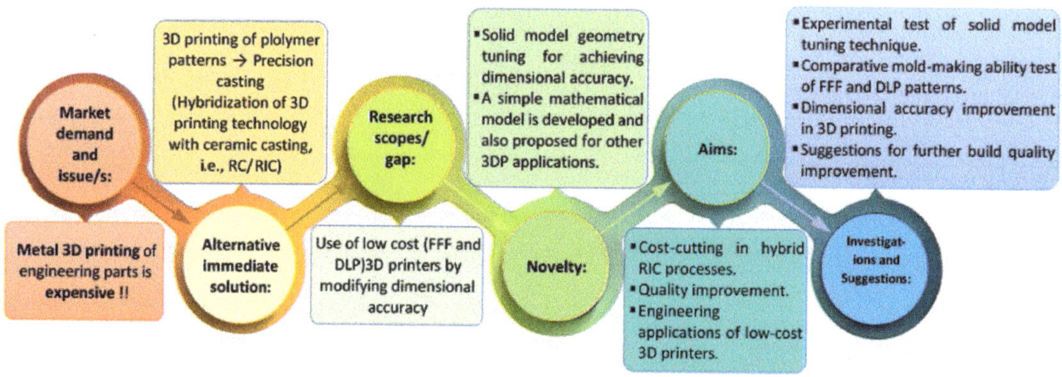

Figure 8. The inherent target of the present investigation of solid model tuning.

1.3. Statement of the Research Objectives

This work aims to explore the potential application of affordable prototype-making 3D printers in engineering product manufacturing through a novel solid model tuning technique and its implementation in rapid casting (RC). To achieve these aims, the following objectives are set in this experimental investigation to improve the dimensional accuracy and quality of the final product.

- To develop a solid model (1st phase) and 3DP by FFF and SLA (DLP) technology and measure the dimensional variations in *X*, *Y*, and *Z* directions.
- Tuning/modifying first solid mode by adjusting dimensional variations (allowances) in *X*, *Y*, and *Z* directions and developing a modified solid model (second) and 3DP for dimensional accuracy improvement verification.
- A modified solid model is used for pattern printing in the FFF and SLA (DLP) process and tested to make the same metallic parts in rapid casting.
- Exploring the technical issues and remedial measure findings as the future scope of this work.

2. Materials and Methods

2.1. Three-Dimensional Printing Machine and Material Details

A budget-friendly DLP and an FFF 3DP machine are used in this investigation. The machine and material details are as follows:

SLA 3D printer: JX215 (China make) DLP (light curing LCD) 3D printer with volume-built rate (215 × 135 × 15) mm^3 per hour, LCD screen spot size 0.0067 mm^2, 405 nm light wavelengths based wax photopolymer resin, layer thickness 0.025 mm. Slicer 2.4.0 software supplied with the machine makes STL files from Auto CAD 3D models.

FFF 3D printer: Ultimaker S3(USA made) FFF 3D printer with built size (230 × 190 × 200) mm^3, speed ~24 mm^3/s, nozzle size 0.25 mm, PLA filament, nozzle temperature 220 °C.

2.2. RC Mould Making and Material Details

Silica sand (180 to 300 microns) and gypsum, mixed in a 50:50 wt.% ratio with water, are stirred to create clay. A small amount of acetone (1 to 2%) is added to reduce bubbles in the clay. The patterns, which are fixed inside the mould box, are submerged in the ceramic slurry and left for an hour. After solidification, the moulds are baked in a muffle furnace at temperatures of 650 °C, 450 °C, and 350 °C for 2 h, 4 h, and 8 h, respectively. Brass is used as the working material for casting tests.

2.3. Techniques for Solid Model Geometry Tuning

The technique and experimental procedure for the tuning of the solid model dimensions are described with the process flow cycle in Figure 9. All the measured dimensions are the average of three consecutive measurements.

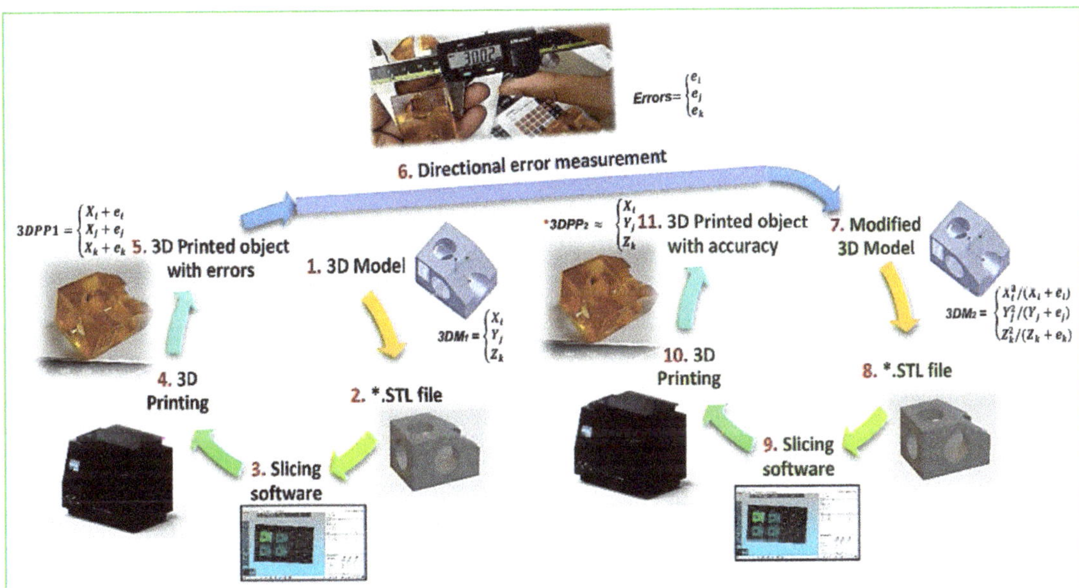

Figure 9. The process cycle of solid model tuning for error compensation in dimensional accuracy improvement.

The Error-Tuning Model

The model for calculating the dimensions for the tuning of the 3D model, aiming to obtain the targeted dimensions, is as follows:

Assume that the 3D model 1 dimension, $3DM_1 = \begin{cases} X_i \\ Y_j \\ Z_k \end{cases};$

which is used to make the 3D-printed parts. But, after 3DP, the measured 3D-printed pattern1 dimensions are

$$3DPP_1 = \begin{cases} X_i + e_i \\ Y_j + e_j \\ Z_k + e_k \end{cases};$$

and now the tuning of the 3D model dimensions is

$$3DM_2 = \begin{cases} X_i^2/(X_i + e_i) \\ Y_j^2/(Y_j + e_j) \\ Z_k^2/(Z_k + e_k) \end{cases};$$

i.e., $3DM_2$ will expectedly produce $3DPP_2$ with the dimensions of $3DM_1$

$$3DPP_2 \approx \begin{cases} X_i \\ Y_j \; ; \\ Z_k \end{cases}$$

so,

$$\underbrace{\begin{cases} X_i \\ Y_j \\ Z_k \end{cases}}_{\text{Final product dimensions}} \rightarrow \underbrace{\begin{cases} X_i^2/(X_i + e_i) \\ Y_j^2/\left(Y_j + e_j\right) \\ Z_k^2/(Z_k + e_k) \end{cases}}_{\text{Tuned 3D model dimensions}}$$

where, X_i, Y_j, and Z_k are the actual 3D model dimensions in the X, Y, and Z directions, respectively, and e_i, e_j, and e_k are the measured errors on 3D-printed parts in the X, Y, and Z directions, respectively, using the same 3D model.

2.4. Rapid Casting Test

The tuned 3D model is employed in 3D printing to generate an enhanced pattern. This pattern is then attached to a wax gating system and placed beneath a layer of ceramic clay, resulting in the solidification of a green ceramic mould, as depicted in Figure 10a. The green ceramic mould is then subjected to baking inside a muffle furnace to sacrifice the wax pattern assembly and form a mould cavity. The worked metal (brass) is melted and heated up to 1100 °C and then poured into the ceramic mould cavity, as shown in Figure 10b.

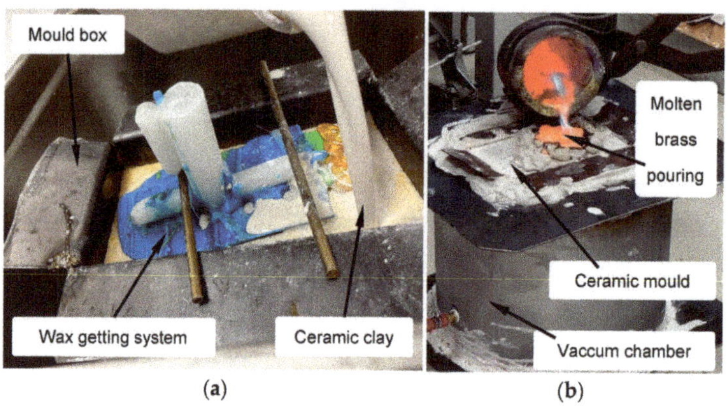

Figure 10. Rapid casting (**a**) green mould making and (**b**) pouring of molten material in a ceramic mould.

After the workpiece solidifies and cools down, the ceramic mould is broken, and the excess impressions (gating system: sprue, runner, risers, etc.) are removed, and the surface is cleaned with a metallic brush and shop water.

3. Results and Discussions

Material extrusion/FFF and SLA/DLP 3D printing technologies are among the recent low-cost 3D printing machines available on the market. The improvement in the dimensional accuracy of these 3D printing processes within engineering tolerances can be effective for engineering applications, potentially reducing the overall investment cost of final products. Such an improvement can encourage the growth of small-scale and micro-industries, as illustrated in Figure 11 below.

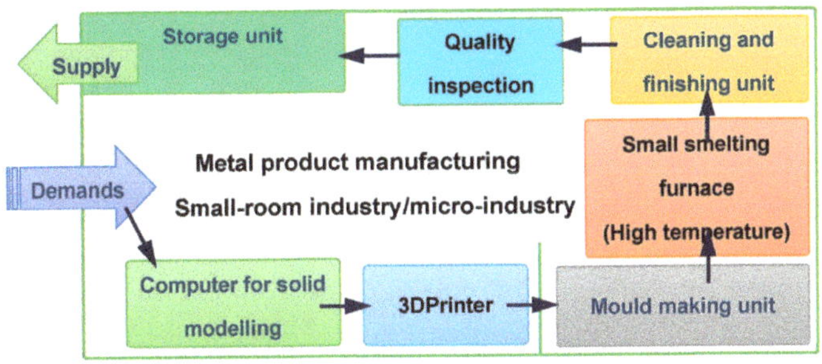

Figure 11. A job shop floor plan for a future RIC micro-industry.

The future RIC industry described above would have the capability to manufacture a wide variety of advanced materials and products for testing and job production.

3.1. Presentation of Empirical Data Showcasing Dimensional Accuracy before and after Geometry Tuning

A simple 3D model is made for the ease of validating the error tuning model, and the measured (as in Figure 12) and calculated data are given in Table 5.

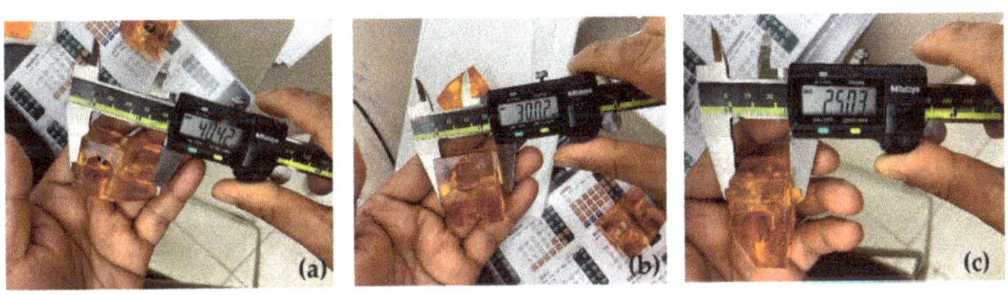

Figure 12. Three-dimensional-printed specimen made from tuned model and error measurement (**a**) in X, (**b**) in Y, and (**c**) in Z directions.

Table 5. Three-dimensional-printed parts dimensions before and after 3D model tuning.

3D Object Directions	1st Model Dimension (Actual Target)	1st Printed Object Dimensions (mm)	Tuned 2nd Model Dimensions (mm)	3DP Specimen after Solid Model Tuning (mm)	Actual Error (%) (Before Validation)	Error (%) in 3D-Printed Specimen after Validation	Reduction in Error (%) after First Validation
X	40	40.48	39.52	40.42	1.2%	1.05%	0.15%
Y	30	30.12	29.88	30.02	0.4%	0.07%	0.33%
Z	25	24.51	25.51	25.03	−2.04%	0.12%	2.16%

For validating the tuned model, the dimensions of the 3D-printed parts are measured before and after 3D model tuning, and the improvement in the dimensional accuracy is given in Table 5.

By applying the developed mathematical model here, the dimensional accuracy of rapid casting could be significantly improved, as observed in Table 5.

3.2. Observations in Rapid Casting

The trial experiment was performed to realize the rapid casting (as in Figure 13) ability using the FFF and SLA(DLP) wax patterns. The observations and results of the rapid casting test are provided in Table 6.

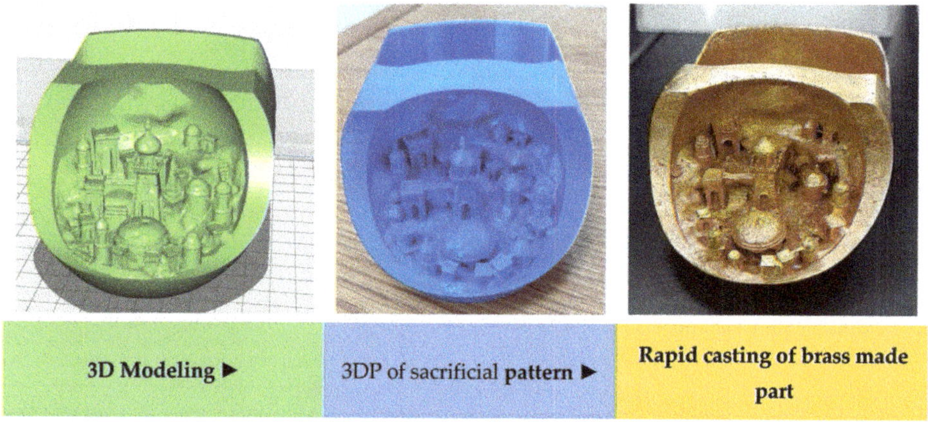

Figure 13. Rapid ceramic casting process sequence in reality in this investigation.

Table 6. Observations on applications of low-cost FFF and DLP 3DP in rapid casting.

Pattern-Making Process and Material	Observations			
	Pattern Surface Quality	Pattern Burning Effects	Mould Quality	Rapid Casting Status
FFF/PLA:	▪ Rough surface (Ra ≈ 20 μm); ▪ Staircasing in slanting surfaces	▪ Smell in PLA burnin ▪ Small ash formed in the mould cavity	▪ Mould formed successfully ▪ Some ash stacked mould on the inner side	▪ Casting completed successfully ▪ Minute surface defects made due to ash ▪ Stair casting impression observed
SLA(DLP)/ Wax:	▪ Smooth surface (Ra ≈ 5 μm); ▪ Staircasing in slanting surfaces	▪ No smell ▪ No ash formed	▪ A large crack formed in the mould due to the thermal expansion of the wax pattern	▪ Failed (due to mould damage)

From the above observation, RC was not successful using the SLA wax pattern, which was generated using the UV photopolymer. The researcher D. Wang et al., in 2018 [35], also reported that photosensitive resin is not suitable for IC. On the other hand, it could be said that the FFF pattern is more effective in making medium-sized ceramic moulds. This might be due to the honeycomb/net structure/vacant space inside the pattern body surfaces that may easily accommodate thermal stress [34], as schematically shown in Figure 14. R. V. Baier et al. [40] demonstrated that low-cost, in-house FFF printers can efficiently produce porous structures comparable to those made by commercial, higher-cost FFF 3D printers.

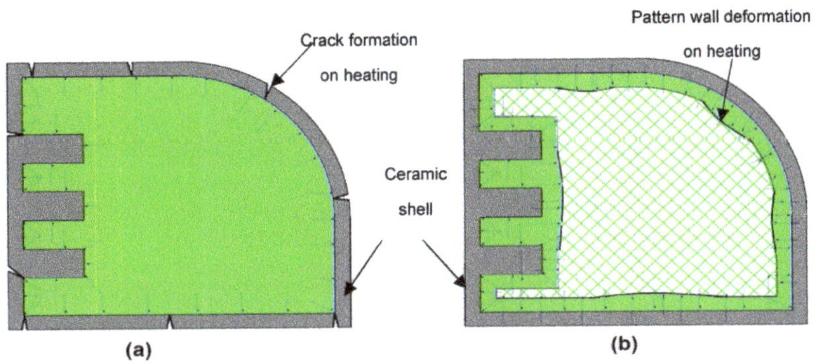

Figure 14. Nature of thermal stress compensation in (**a**) solid/inert pattern by SLA and (**b**) honeycomb/net structured pattern by FFF method and their effect on ceramic shell.

However, the rough surface of the FFF pattern and ash PLA content hinder the RC product's quality. So, FFF wax filament (non-ash content and no smell in burning) may be selected for pattern 3DP with a surface modification for RC. The FFF pattern can save pattern material and it is easier to modify the surface to improve the surface quality.

A circular nozzle is used to extrude the filament for layer deposition, which is a common practice. Surface roughness, height loss, gap between the layers, etc., hinder the quality of FFF 3DP. To improve surface quality in FFF 3DP parts/patterns, chemical treatment, vapour treatment, polishing [34], laser polishing [41], and micro-grit/sand blasting [42] techniques are practised by some researchers. Laser polishing shows promise in enhancing strength and sustainability. Additionally, the printing quality of low-cost 3D printers can be improved by increasing machine stiffness and reducing chatter [43].

The present author proposes using a rectangular FFF nozzle with rounded corners to extrude a filleted square cross-section filament for layer deposition, as illustrated in Figure 15a. This approach is expected to mitigate internal gaps and reduce height loss compared to circular filament deposition, as shown in Figure 15b.

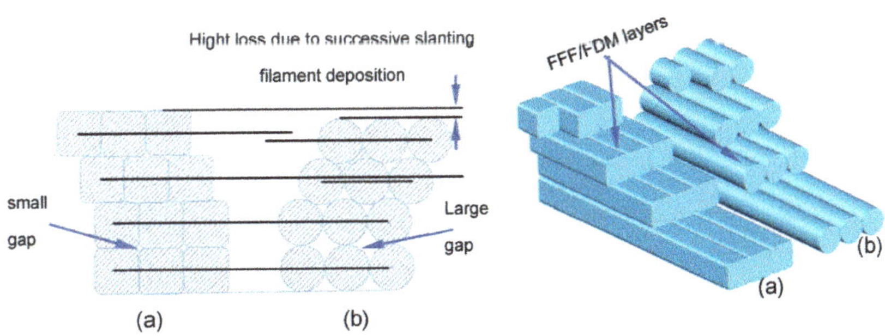

Figure 15. Geometrically proven improvement in deposition quality in FFF 3DP using (**a**) filleted square cross-section compared to (**b**) round cross-section filament.

It is clear from the above figure that the filleted square cross-section extrusion nozzle may have an effect on reducing the internal gaps in FFF-made parts, which could be a future area of investigation.

4. Conclusions

The research survey and experimental investigation conducted in this study aim to develop and establish solid model tuning for improving dimensional accuracy and implementing it in engineering and industrial applications. This work leads to the following observations being made:

- Highly complex parts that cannot be manufactured using other manufacturing processes can be effectively produced through direct metal 3D printing. For low- to medium-complexity metal parts, cost-effective production can be achieved through the RC/RIC process.
- A substantial amount of research and investigation is crucial for the commercialisation of the RC/RIC process.
- Affordable 3D printing (3DP) technologies, such as FFF or DLP processes, can potentially be leveraged for cost-effective engineering applications and the manufacturing of metal products using the RC/RIC route in job shop production houses through solid model tuning.
- The proposed and experimentally tested mathematical model for solid model geometry tuning can be applied to other 3D printing methods to enhance the dimensional accuracy of 3D-printed products, making them compliant with engineering tolerances and suitable for various applications.
- This study examines the ceramic mould-building capacity for rapid casting (RC) using medium-sized (4 to 10 cm) wax-made DLP patterns and PLA-made FFF patterns. The hollow interior of the FFF-based 3D-printed patterns facilitates successful RC.
- The suggested filleted square shape extruder nozzle may deposit better build quality products than the commonly used circular extruder in FFF 3DP.
- Such low-cost process development could enable the establishment of a room-based or micro-scale RC/RIC industry with a low initial investment cost.

Immense research scopes exist in this domain to improve product quality and minimize production costs. Hence, rapid casting is a solution to produce low- to medium-complexity metallic products in job-to-batch production.

Funding: The author extends his appreciation to the Deanship of Scientific Research, Imam Mohammad Ibn Saud Islamic University (IMSIU), Saudi Arabia, for funding this research work through Grant No. (221414006).

Data Availability Statement: Open to use.

Conflicts of Interest: The authors declare no conflict of interest.

Nomenclature

3DP	3D Printing
AI	Artificial Intelligence
AM	Additive Manufacturing
DLP	Digital Light Processing
DLS	Digital Light Synthesis
DM	Direct Manufacturing
DMLS	Direct Metal Laser Sintering
DMP	Direct Metal Printing
EBM	Electron Beam Melting
FDM	Fused Deposition Modelling
FEP	Fluorinated Ethylene Propylene
FFF	Fused Filament Fabrication
IC	Investment Casting
LPBF	Laser Powder Bed Fusion
MJF	Multi Jet Fusion
MMC	Metal Matrix Composite
PLA	Polylactic Acid
RC	Rapid Casting
RP	Rapid Prototyping
SLA	Stereolithography
SLM	Selective Laser Melting
SLS	Selective Laser Sintering

References

1. Cheah, C.M.; Chua, C.K.; Lee, C.W.; Feng, C.; Totong, K. Rapid prototyping and tooling techniques: A review of applications for rapid investment casting. *Int. J. Adv. Manuf. Technol.* **2005**, *25*, 308–320. [CrossRef]
2. Gibson, I.; Rosen, D.; Stucker, B. *Additive Manufacturing Technologies: 3D Printing, Rapid Prototyping, and Direct Digital Manufacturing*, 2nd ed.; Springer: Berlin/Heidelberg, Germany, 2015; ISBN 9781493921126.
3. Vafadar, A.; Guzzomi, F.; Rassau, A.; Hayward, K. Advances in metal additive manufacturing: A review of common processes, industrial applications, and current challenges. *Appl. Sci.* **2021**, *11*, 1213. [CrossRef]
4. Ngo, T.D.; Kashani, A.; Imbalzano, G.; Nguyen, K.T.Q.; Hui, D. Additive manufacturing (3D printing): A review of materials, methods, applications and challenges. *Compos. Part B Eng.* **2018**, *143*, 172–196. [CrossRef]
5. Singh, J.; Singh, R.; Singh, H. Dimensional accuracy and surface finish of biomedical implant fabricated as rapid investment casting for small to medium quantity production. *J. Manuf. Process.* **2017**, *25*, 201–211. [CrossRef]
6. Niemelä, M.; Shi, A.; Shirowzhan, S.; Sepasgozar, S.M.E.; Liu, C. 3D printing architectural freeform elements: Challenges and opportunities in manufacturing for industry 4.0. In Proceedings of the 2019 36th International Symposium on Automation and Robotics in Construction (ISARC 2019), Banff, AB, Canada, 21–24 May 2019; pp. 1298–1304. [CrossRef]
7. Varotsis, A.B. 3D Printing Trends Q1 2019. Available online: https://www.hubs.com/blog/3d-printing-trends-q1-2019/ (accessed on 9 February 2023).
8. Roberts, T.; Bartkova, B. Additive Manufacturing Trend Report 2021 (3D Printing Market Growth in the Year of the COVID-19). Available online: https://f.hubspotusercontent10.net/hubfs/4075618/Additive%20manufacturing%20trend%20report%202021.pdf?utm_campaign=Gated%20Content%20Downloads&utm_medium=email&_hsmi=82605589&_hsenc=p2ANqtz-9pM5X9Hu5jL_49jLKLdBARaGb5PVPhM3kYwzk2c36z2bmdpMZRLJiSd4n5ZKSZOJoCbPnSwFXfi64pLJvjvFjceDPf3w&utm_content=82605589&utm_source=hs_automation (accessed on 12 August 2023).
9. 3D Printing Market Size. Available online: https://www.fortunebusinessinsights.com/industry-reports/3d-printing-market-101902 (accessed on 10 September 2023).
10. Industrial 3D Printing Market Size, Share and Growth. Available online: https://www.alliedmarketresearch.com/industrial-3d-printing-market-A17129 (accessed on 10 September 2023).
11. 3D Printing Market Size to Worth around USD 98.31 BN. Available online: https://www.globenewswire.com/en/news-release/2023/03/28/2635882/0/en/3D-Printing-Market-Size-to-Worth-Around-USD-98-31-BN-by-2032.html (accessed on 10 August 2023).
12. MIT Comparing AM Processes. Available online: http://apt.mit.edu/am-process-comparisons (accessed on 1 February 2022).
13. Alsoufi, M.S.; Elsayed, A.E. Surface Roughness Quality and Dimensional Accuracy—A Comprehensive Analysis of 100% Infill Printed Parts Fabricated by a Personal/Desktop Cost-Effective FDM 3D Printer. *Mater. Sci. Appl.* **2018**, *9*, 11–40. [CrossRef]
14. Pandian, A.; Belavek, C. A review of recent trends and challenges in 3D printing. In Proceedings of the 2016 ASEE North Central Section Conference, Michigan, MI, USA, 18–19 March 2016; pp. 1–17.
15. Yarbrough, J.; Shearer, A.B.; Bent, S.F. Next generation nanopatterning using small molecule inhibitors for area-selective atomic layer deposition. *J. Vac. Sci. Technol. A* **2021**, *39*, 021002. [CrossRef]
16. How to get all-metal parts with 3D FDM printing. Available online: https://filament2print.com/gb/blog/70_sintering-metal-parts-fdm-3d-printing.html (accessed on 12 August 2023).
17. Markforged Types of 3D Printing in Metal. Available online: https://markforged.com/resources/learn/design-for-additive-manufacturing-metals/metal-additive-manufacturing-introduction/types-of-3d-printing-metal (accessed on 18 September 2023).
18. Wohlers Associates. Wohlers Report 2021. 2021. Available online: wohlersassociates.com (accessed on 10 September 2023).
19. Stefaniak, A.B.; Bowers, L.N.; Knepp, A.K.; Luxton, T.P.; Peloquin, D.M.; Ham, J.E.; Wells, J.R.; Johnson, A.R.; Lebouf, R.F.; Martin, S.B.; et al. Particle and vapor emissions from vat polymerization desktop-scale 3-dimensional printers. *J. Occup. Environ. Hyg.* **2020**, *16*, 519–531. [CrossRef] [PubMed]
20. Formlabs Introduction to Casting for 3D Printed Jewelry Patterns. Available online: https://formlabs.com/company/recommended-casting-houses (accessed on 10 August 2023).
21. Yap, Y.L.; Yeong, W.Y. Additive manufacture of fashion and jewellery products: A mini review: This paper provides an insight into the future of 3D printing industries for fashion and jewellery products. *Virtual Phys. Prototyp.* **2014**, *9*, 195–201. [CrossRef]
22. Ott, D.; Raub, C.J. Investment Casting of Gold Jewellery. *Gold Bull.* **1986**, *19*, 34–39. [CrossRef]
23. Pattnaik, S.; Jha, P.K.; Karunakar, D.B. A review of rapid prototyping integrated investment casting processes. *Proc. Inst. Mech. Eng. Part L J. Mater. Des. Appl.* **2014**, *228*, 249–277. [CrossRef]
24. A Modern Casting Staff Report. Census of World Casting Production: Total Casting Tons Dip in 2019; Modern Casting, 2021, 28. Available online: https://www.qgdigitalpublishing.com/publication/?m=55001&i=687958&p=30&ver=html5 (accessed on 12 August 2023).
25. Pattnaik, S.; Karunakar, D.B.; Jha, P.K. Developments in investment casting process—A review. *J. Mater. Process. Technol.* **2012**, *212*, 2332–2348. [CrossRef]
26. Stefanescu, D.M. *ASM Handbook Volume 15:Casting*; ASM International: Michigan, MI, USA, 1998; ISBN 978-0871700223.
27. Olson, D. What Is Investment Casting and How Does It Work? Available online: https://www.metaltek.com/blog/what-is-investment-casting-and-how-does-it-work/ (accessed on 19 September 2023).

28. The Accuracy of Investment Casting. Available online: https://www.improprecision.com/introduction-investment-casting-accuracy-grades/ (accessed on 10 September 2023).
29. Xu, M.; Lekakh, S.N.; Von Richards, L. Thermal property database for investment casting shells. *Int. J. Met.* **2016**, *10*, 342–347. [CrossRef]
30. Chen, Y.; Zhao, E.; Kong, F.; Xiao, S. Fabrication of thin-walled high-temperature titanium alloy component by investment casting. *Mater. Manuf. Process.* **2013**, *28*, 605–609. [CrossRef]
31. Klotz, U.E.; Legner, C.; Bulling, F.; Freitag, L.; Faßauer, C.; Schafföner, S.; Aneziris, C.G. Investment casting of titanium alloys with calcium zirconate moulds and crucibles. *Int. J. Adv. Manuf. Technol.* **2019**, *103*, 343–353. [CrossRef]
32. Li, F.; Ji, X.; Wu, Z.; Qi, C.; Lai, J.; Xian, Q.; Sun, B. Digital light processing 3D printing of ceramic shell for precision casting. *Mater. Lett.* **2020**, *276*, 2–5. [CrossRef]
33. Mukhtarkhanov, M.; Perveen, A.; Talamona, D. Application of stereolithography based 3D printing technology in investment casting. *Micromachines* **2020**, *11*, 946. [CrossRef]
34. Tiwary, V.K.; Arunkumar, P.; Deshpande, A.S.; Rangaswamy, N. Surface enhancement of FDM patterns to be used in rapid investment casting for making medical implants. *Rapid Prototyp. J.* **2019**, *25*, 904–914. [CrossRef]
35. Wang, D.; Dong, A.; Zhu, G.; Shu, D.; Sun, J.; Li, F.; Sun, B. Rapid casting of complex impeller based on 3D printing wax pattern and simulation optimization. *Int. J. Adv. Manuf. Technol.* **2018**, *100*, 2629–2635. [CrossRef]
36. Macků, M.; Horáček, M. Applying RP-FDM Technology to Produce Prototype Castings Using the Investment Casting Method. *Arch. Foundry Eng.* **2012**, *12*, 75–82. [CrossRef]
37. Wang, S.; Miranda, A.G.; Shih, C. A study of investment casting with plastic patterns. *Mater. Manuf. Process.* **2010**, *25*, 1482–1488. [CrossRef]
38. Vaezi, M.; Safaeian, D.; Shakeri, M. Integration of reverse engineering and rapid technologies for rapid investment casting of gas turbine blades. *Virtual Phys. Prototyp.* **2011**, *6*, 225–239. [CrossRef]
39. Bassoli, E.; Gatto, A.; Iuliano, L.; Violante, M.G. 3D printing technique applied to rapid casting. *Rapid Prototyp. J.* **2007**, *13*, 148–155. [CrossRef]
40. Vallejos Baier, R.; Contreras Raggio, J.I.; Toro Arancibia, C.; Bustamante, M.; Pérez, L.; Burda, I.; Aiyangar, A.; Vivanco, J.F. Structure-function assessment of 3D-printed porous scaffolds by a low-cost/open source fused filament fabrication printer. *Mater. Sci. Eng. C* **2021**, *123*, 111945. [CrossRef]
41. Mushtaq, R.T.; Iqbal, A.; Wang, Y.; Khan, A.M.; Petra, M.I. Advancing PLA 3D Printing with Laser Polishing: Improving Mechanical Strength, Sustainability, and Surface Quality. *Crystals* **2023**, *13*, 626. [CrossRef]
42. Minetola, P.; Calignano, F.; Galati, M. Comparing geometric tolerance capabilities of additive manufacturing systems for polymers. *Addit. Manuf.* **2020**, *32*, 101103. [CrossRef]
43. Minetola, P.; Galati, M. A challenge for enhancing the dimensional accuracy of a low-cost 3D printer by means of self-replicated parts. *Addit. Manuf.* **2018**, *22*, 256–264. [CrossRef]

machines

MDPI

Article

Experimental Study and 3D Optimization of Small-Scale Solar-Powered Radial Turbine Using 3D Printing Technology

Ahmed M. Daabo [1,*], Ali Abdelhafeez Hassan [2,*], Muhammad Anser Bashir [3,4], Hudhaifa Hamza [1], Shahad Salim [1], Aisha Koprulu [5], Tawfik Badawy [6], Saad Mahmoud [7] and Raya Al-Dadah [7]

1 Mining Engineering Department, College of Petroleum and Mining Engineering, The University of Mosul, Mosul 41200, Iraq
2 School of Computing, Engineering and Digital Technologies, Teesside University, Middlesbrough TS1 3BX, UK
3 Department of Engineering, Roma Tre University, Via Della Vasca Navale 79, 00146 Rome, Italy; anser.me@must.edu.pk
4 Department of Mechanical Engineering, Mirpur University of Science & Technology (MUST), Mirpur 10250, AJK, Pakistan
5 Aeronautical Technical Engineering Department, Technical College, Al-Kitab University, Kirkuk 36001, Iraq
6 Mechanical Power Engineering Department, Cairo University, Giza 12613, Egypt
7 School of Engineering, The University of Birmingham, Edgbaston, Birmingham B15 2TT, UK
* Correspondence: ahmeddaboo@uomosul.edu.iq (A.M.D.); a.hassan@tees.ac.uk (A.A.H.)

Abstract: Small-Scale Turbines (SSTs) are among the most important energy-extraction-enabling technologies in domestic power production systems. However, owing to centrifugal forces, the high rotating speed of SSTs causes excessive strains in the aerofoil portions of the turbine blades. In this paper, a structural performance analysis is provided by combining Finite Element Methods (FEM) with Computational Fluid Dynamics (CFD). The primary objective was to examine the mechanical stresses of a Small-Scale Radial Turbine (SSRT) constructed utilizing 3D printing technology and a novel plastic material, RGD 525, to construct a SSRT model experimentally. After introducing a suitable turbine aerodynamics model, the turbine assembly and related loads were translated to a structural model. Subsequently, a structural analysis was conducted under various loading situations to determine the influence of different rotational speed values and blade shapes on the stress distribution and displacement. Maximum von Mises and maximum main stresses are significantly affected by both the rotor rotational speed and the working fluid input temperature, according to the findings of this research. The maximum permitted deformation, on the other hand, was more influenced by rotational speed, while the maximum allowable fatigue life was more influenced by rotating speed and fluid intake temperature. Also, the region of the tip shroud in the rotor had greater deflection values of 21% of the blade tip width.

Keywords: small-scale turbine; FEA; 3D printing; stress distribution; RGD 525

1. Introduction

The increasing request for electric power and, at the same time, the strict environmental restrictions on thermal pollution have increased the necessity for clean energy systems [1–3]. Small-scale turbines in Solar-Powered Brayton Cycle SPBC application offer a wide range of out power values suitable for relatively small scales systems to large scales [4]. Several authors [5,6] investigated in detail components like the thermal receiver and optimized the cycle performance, but they neglected the turbines' performance. Others [7,8] studied the cycle effect of wake on the performance of turbines. Aerodynamically, several researchers [9,10] have improved the performance of SST; however, they neglected the structural analysis. Also, an attempt to aerodynamically optimize the rotor of an impulse axial turbine for oscillating water column was achieved by Gomes et al. [11].

Recently some studies regarding the structural analysis were conducted by some other researchers. Shanechi et al. [12] presented a CFD-FEM-coupled analysis of a relatively high pressure ratio radial inflow turbine where it was designed with some emphasis on blades geometry to enhance a turbine's output power and efficiency. Their FEM analysis focused on blade enhancements in terms of its stresses and deformation. Barsi et al. [13] investigated enhancing the aerodynamic performance and reduction in thermal stresses of a micro gas radial turbine employing a multidisciplinary optimization technique. The camber and half thickness of the rotor blade were optimized to improve turbine efficiency by about 5% while maintaining the allowable material stresses level. Nevertheless, mechanical fatigue and modal analysis were not included in the aforementioned research, as well as the boundary conditions range and output power values being dissimilar to the current study.

Similar research efforts have been carried out [14–17]. Both mechanical and thermal loads, with their impact on the aerodynamic and deformation performance of the impeller for a micro gas radial turbine, have been analysed [14]. It was shown that the maximum deformation was achieved at the blade inducer tip, and that the aerodynamic performance and structural reliability of the impeller blade tip were degraded by its fluctuated deformation. Optimizing the profile of the impeller hub section reduced the deformation magnitude of the impeller; nonetheless, the aerodynamic performance was not under the researchers' spotlight. Similarly, Feng et al. [17] enhanced the strength of the impeller hub by optimizing the blade shape of a 100 kW microturbine. It was shown that the aerodynamic and structural performances were affected by the inlet relative flow incident angle of the impeller with an optimum angle of $-32°$. The optimization of the micro-radial turbine was achieved using the inverse design method (via CFD and FEM) of two key parameters—the blade profile and the blade thickness—which resulted in in an effective improvement in the blade strength without a noticeable decrease in the turbine efficiency. Fu et al. [15] evaluated improving the aerodynamic performance, strength and weight of the wheel design of radial turbines, utilizing an integrated type of optimizing method. The results indicated that turbine wheel weight was decreased by about 50% and, at the same time, a satisfactory stress distribution and good aerodynamic performance were realized.

Aside from the current work linking the mean line, the 3D aerodynamic CFX, the structural and the thermal analyses, it also uses 3D printing technology to experimentally manufacture the designed SSRT with RGD 525 material. It is high-temperature opaque brittle plastic material which has been recently used in 3D printer technology, and as a model or support material as well [18]. Usually, this material needs to be treated thermally, following a specific procedure [19], in order to improve its properties by alternating it from orthotropic to isotropic material.

As a result, this work aims to optimize the performance of SSRT by taking into account all of the aforementioned concerns and determining the SSRT geometry. The 1D Vista RTD (part of ANSYS 2015) was used to determine the beginning dimensions and performance of each type for the turbine preliminary design. After that, it was combined with ANSYS CFX [20] to determine the 3D turbine shape and assess its performance. The baseline design was originally optimized utilizing the same authors' prior work's ANSYS®15 Design Exploration programme for 3D optimization purposes based on evolutionary algorithms. Following the achievement of an aerodynamically optimum design, structural calculations were carried out to determine the best structural specifications.

2. Methodology and Experimental Work

In the authors' previous works [21,22], the aerodynamically optimum SSRT was achieved and, in the current work, the structural analysis will be further analysed to ensure a robust optimized turbine with a good performance for the required application. The optimum design point for both the stator and the rotor was extracted as a complete model and, then, by using the integrated CFX-FE Workbench in ANSYS®18, that optimum design was transferred in order to assess the stress values, the thermal analysis and the deflection and the fatigue analysis.

Figure 1 expresses the followed procedure during the current analysis. From this figure, it can be noticed that the object of this research is to concentrate on the last seven sets of the blocks as the first seven sets were already achieved in the authors' previous publications regarding the SSRT. Once that was achieved, the complete design of all other relevant parts was established using CAD software SOLIDWORKS in order to include all of them in both the thermal and the structural analyses for the RGD 525 material. The complete model of the SSRT is given in Figure 2.

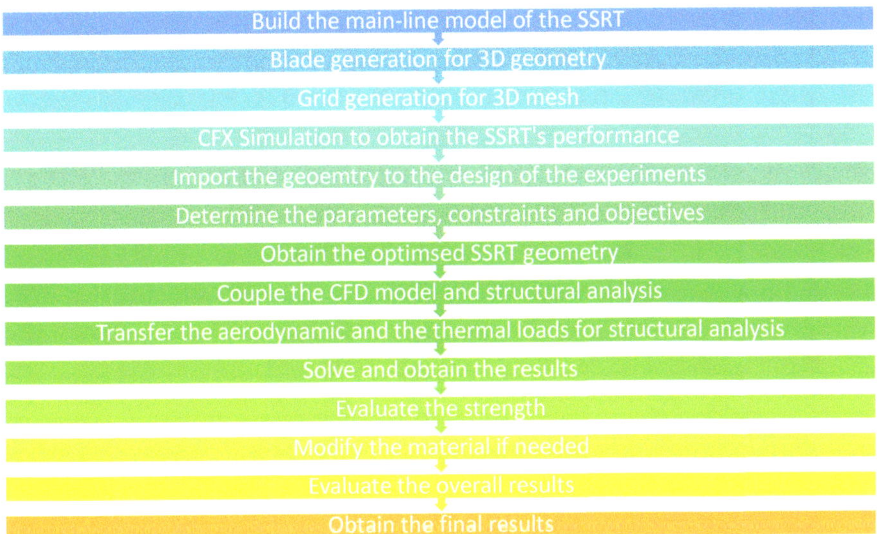

Figure 1. Overview process of the analyses' procedure.

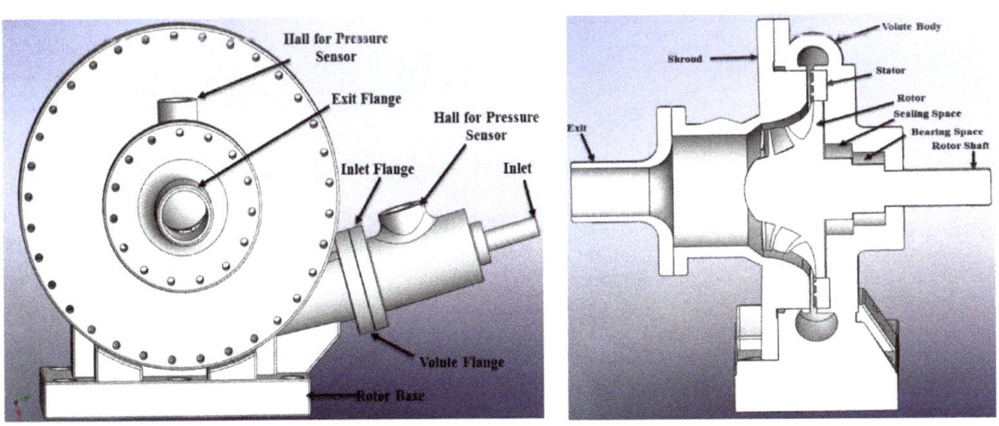

Figure 2. The complete model of the designed SSRT (**left**) and its cross-sectional view (**right**).

Using a programmable oven, and based on the requirement mentioned by the material manufacturer [19], a thermal treatment for the RGD 525 was carried out using thermal furnaces according to a specific cycle, as detailed in Figure 3. The tensile test results after the material had been thermally treated are shown in Figure 4. The tensile test results improved by roughly 37.5% at 35 °C, as can be seen in the first figure.

(A)

(B)

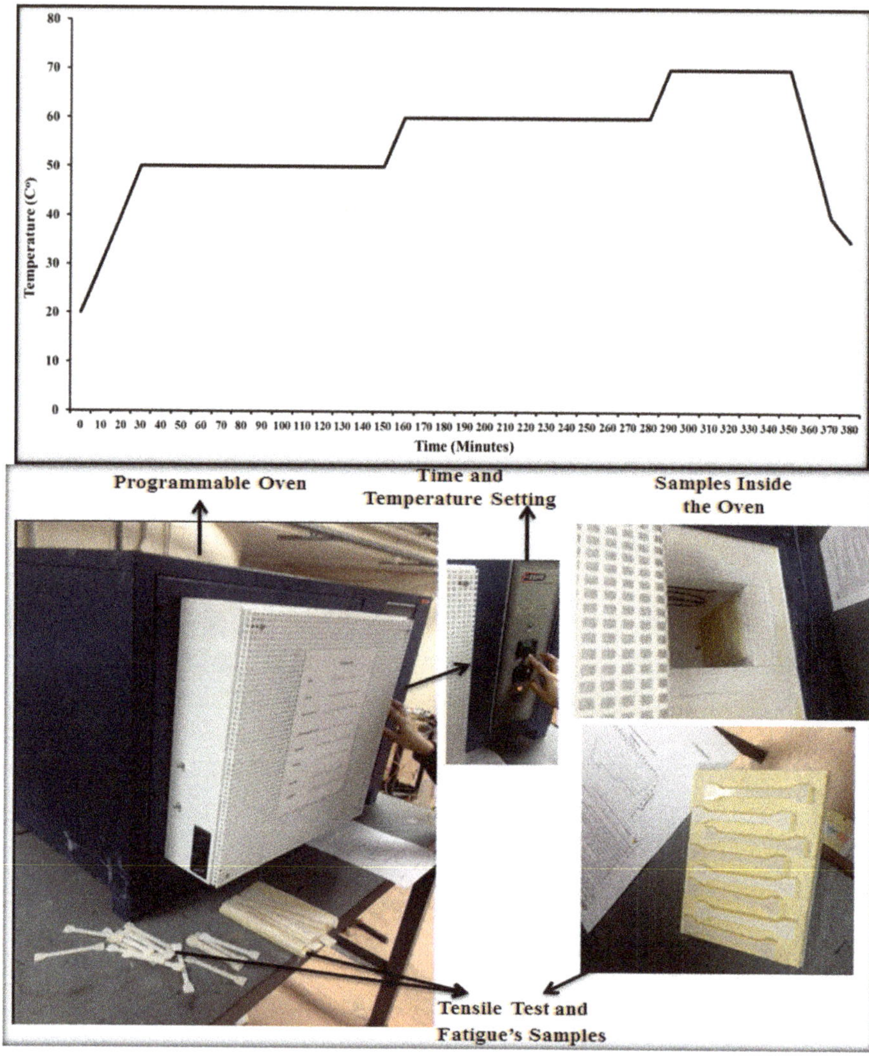

Figure 3. Oven temperature over time (**A**) and (**B**) oven used in the heat treatment.

Fatigue tests were conducted using a servo hydraulic fatigue testing apparatus (INSTRON 8801) that was fitted with self-aligning hydraulic grips. The purpose of this test was to determine the material properties in relation to fatigue analysis, as depicted in Figure 5. The applied load exhibited a cyclic sine wave pattern with a loading ratio of 0.1 at a frequency of 3 Hz. The maximum load for each sample varied between 1.733 kN and 0.473 kN. Figure 6 displays the SSRT components that were produced using StraSys Alaris30 3D printer (PolyJet technology, Holtzman St. Science Park, Israel) at 30 μm layer thickness.

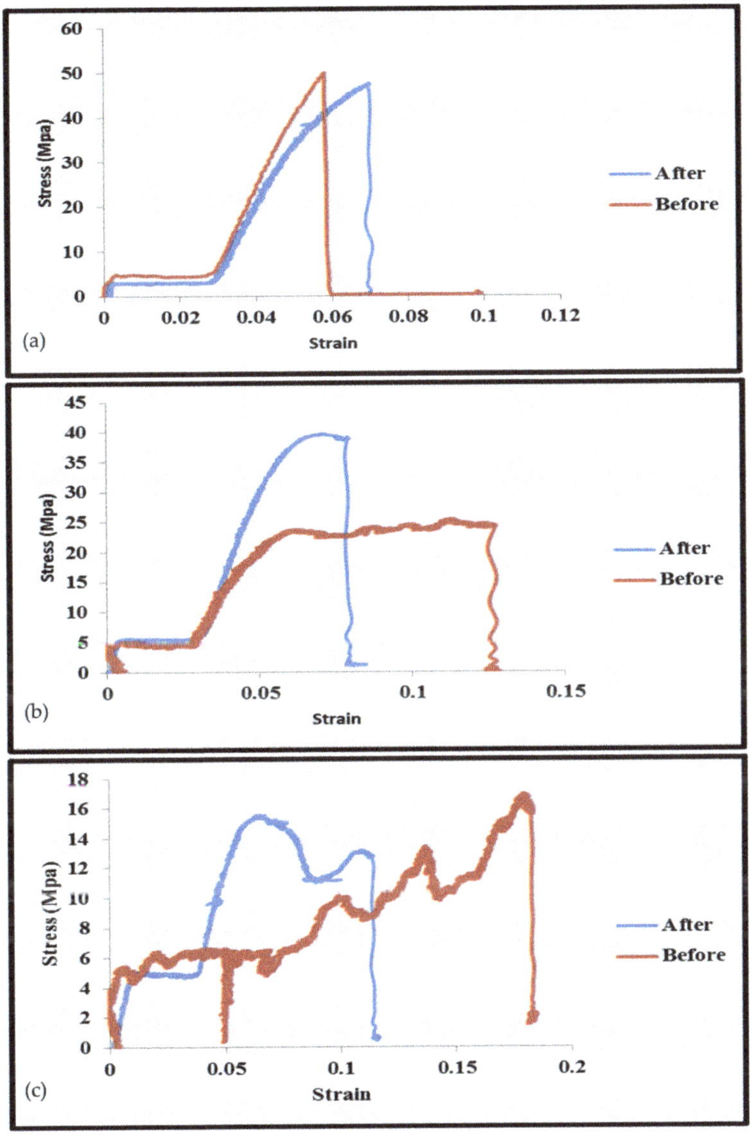

Figure 4. Tensile test results of the treated RGD 525; (**a**) 25 °C, (**b**) 35 °C and (**c**) 45 °C.

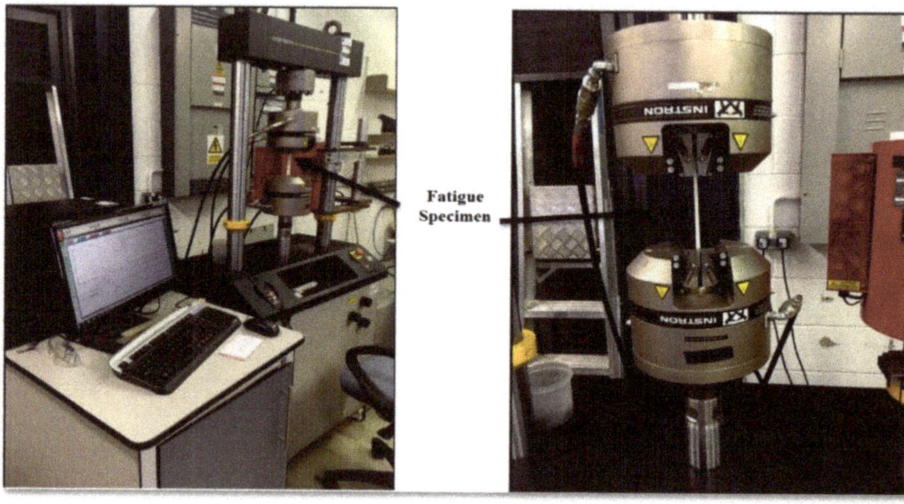

Figure 5. Fatigue test carried out using ISTRON 8801 machine.

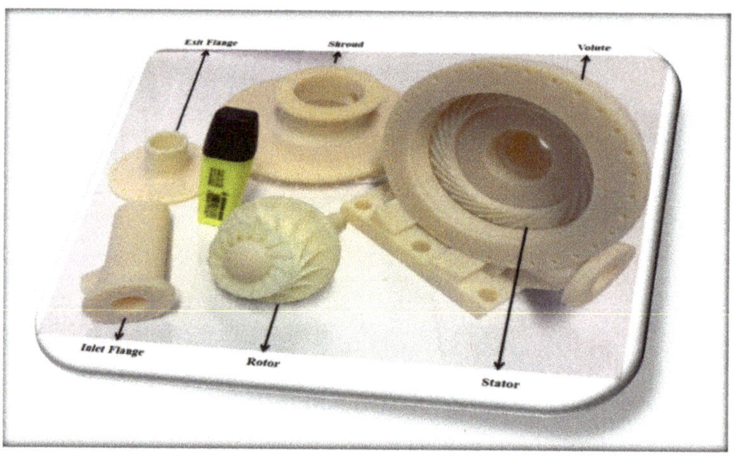

Figure 6. SSRT parts which are printed by 3D printing technology.

3. Governing Equations for Structural Analysis

3.1. Mechanical Stresses Analysis

The stresses that occur in the turbine structure are, in fact, a combination of the thermal gradient effect, static stresses, results from the fluid pressure and vibratory stress, initially from the centrifugal force as a result of rotor rotational velocity. The thermal stress (σ) can be analysed as follows:

$$\sigma = D \, . \, \varepsilon \tag{1}$$

The assessed material is anisotropic (see Table 1) and the elastic stress–strain relations can be written in Cartesian coordinates as follows:

$$\varepsilon_x = \frac{1}{E}\left[\sigma_x - \nu_p\left(\sigma_y + \sigma_z\right)\right] + \alpha\Delta\mathrm{T}(x, y, z) \tag{2}$$

$$\varepsilon_y = \frac{1}{E}\left[\sigma_y - \nu_p\left(\sigma_x + \sigma_z\right)\right] + \alpha\Delta\mathrm{T}(x, y, z) \tag{3}$$

$$\varepsilon_z = \frac{1}{E}\left[\sigma_z - \nu_p\left(\sigma_x + \sigma_y\right)\right] + \alpha\Delta\mathrm{T}(x, y, z) \tag{4}$$

The temperature gradient at a specific point, denoted by (x, y, z), is represented by $\Delta\mathrm{T}(x, y, z)$. The temperature fields were interpolated directly using the computational fluid dynamics (CFD) results.

The steady mechanical stresses are generated by the fluid pressure acting on the blades and the centrifugal force exerted on the rotor. An investigation of the alternating stresses arising from turbulent fluid flow and resonance phenomena [21] is beyond the scope of this study.

The centrifugal force (F_{cf}) can be calculated as follows:

$$F_{cf} = mr{\omega_s}^2 \tag{5}$$

Nevertheless, the determination of the precise magnitude of the centrifugal force was accomplished by taking into account a minute component of the blade section and subsequently integrating it to encompass the entire blade structure, as depicted in the illustration provided.

$$df_{cf} = dm.{\omega_s}^2(R_r + z) \tag{6}$$

where;

$$dm = \rho.A(z)dz \tag{7}$$

$$df_{cf} = \rho{\omega_s}^2.A(z).(R_r + z)dz \tag{8}$$

$$F_{cf}(x) = \int_x^{lb} \rho.{\omega_s}^2.A(z).(R_r + Z)dz \tag{9}$$

By considering the blade as a cantilever with variable cross section area [22], this leads to:

$$\left(\frac{A_{(z)}}{A_r}\right)^{lb} = \left(\frac{A_t}{A_r}\right)^{z} \tag{10}$$

$$A_{(z)} = A_r.\left(\frac{A_t}{A_r}\right)^{z/lb} \tag{11}$$

The equation provided below can be utilized to ascertain the centrifugal force at any given location on the blade:

$$F_{cf(x)} = \rho{\omega_s}^2\left[\frac{A_r.\left(\frac{A_t}{A_r}\right)^{\frac{z}{lb}}.R_r.lb}{\ln\left(\frac{A_t}{A_r}\right)} + \frac{A_r.\left(\frac{A_t}{A_r}\right)^{\frac{z}{lb}}.z.lb}{\ln\left(\frac{A_t}{A_r}\right)} - \frac{A_r.\left(\frac{A_t}{A_r}\right)^{\frac{z}{lb}}.lb^2}{\left[\ln\left(\frac{A_t}{A_r}\right)\right]^2}\right]_x^{lb} \tag{12}$$

Table 1. RGD525 material properties.

Properties	Value
Young's Modulus, MPa	2000
Poisson's Ratio	0.3
Flexural Modulus, MPa	3300
Tensile Strength, MPa	75
Density, kg/m^3	1175
Coefficient of Thermal Expansion, /C	0.000012
Specific Heat, J/kg °C	434
Thermal Conductivity, W/mm °C	0.0605

3.2. Fatigue Analysis

Fatigue is among the most significant phenomenon that must be carefully evaluated in any rotating component. It refers to the failure of a material due to the repeated application of a load, even though the load is less than the material's strength. In this part, a comprehensive assessment for the fatigue study of the SSRT utilizing the FEM programme ANSYS 18 was developed.

Figure 7 depicts a continuous, totally reversed loading cycle which is typically used for fatigue testing. Equation (13) depicts the Goodman relation that governs the endurance limit of a material as a function of the alternating stress, mean stress and ultimate tensile strength of the material. The fatigue loading cycle may be described in a variety of ways, with the most frequent being the usage of the alternating stress value σ_{Alt} and the stress ratio $R = \sigma_{Min}/\sigma_{Max}$.

$$\frac{\sigma_{Alt}}{\sigma_e} + \frac{\sigma_{Mean}}{\sigma_u} = 1 \quad (13)$$

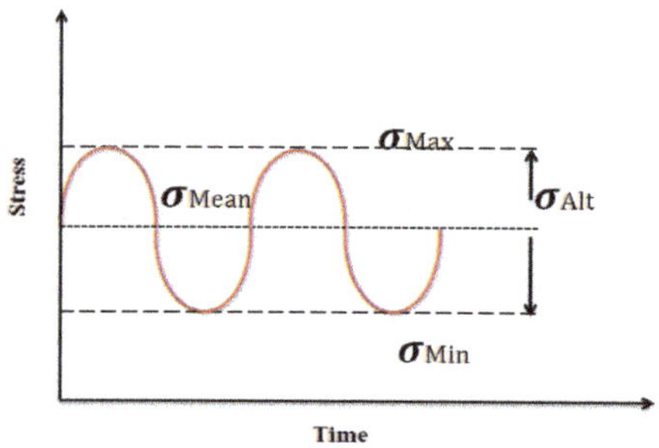

Figure 7. Simultaneous mean and cyclic loading.

However, the relation between the maximum stress used in fatigue loading and the number of cycles to failure can be described as follows [23,24]:

$$\sigma_{Max}\sqrt{\frac{1-R}{2}} = \sigma_f{}'\left(2N_f\right)^b \quad (14)$$

For temperatures (T) different from room temperature T_0, where the fatigue testing is carried out, the maximum stress can be scaled for different temperatures based on the ultimate strength ratio at the associated temperatures as follows [25]:

$$[\sigma_{Max}]_{@T} = [\sigma_{Max}]_{@T_0} \cdot \frac{[\sigma_u]_{@T}}{[\sigma_u]_{@T_0}} \quad (15)$$

Equation (15) is in agreement with the data detailed in reference [26], where the ratio of tensile strength at different temperatures is almost equal to the ratio of fatigue limits at the same temperatures for ABS material.

The fatigue damage for the material can be estimated using the following equation:

$$D_f = \frac{n_i}{N_f} \quad (16)$$

To setup the model, the defining of the new material RGD 525 and the characterizing of the S-N curve obtained from experimental fatigue test were carried out.

4. Numerical Modelling

The structural, fatigue and modal analysis, together with the overall design assessment for the rotor part of the SSRT, were correspondingly carried out, as shown next.

4.1. CFD and Structural Modeling

The design of the blade's shape was created utilizing the Blade-Gen software, and subsequently meshed using the Turbo-Grid module found in ANSYS/Workbench. The structured 3D mesh generation for the complete (stator and rotor) 3D model of the small-scale radial turbine is shown in Figure 8. It should be noted that the grid was precisely refined in the vicinity of the blade's surface to accurately simulate the fluid behaviour, including the viscous effects. The sensitivity analysis of the grid was based on the turbine efficiency and is shown in Figure 9. The results show that there is no change in the predicted efficiency with an increase in the number of elements above 75,000. This indicates that the mesh is adequately fine-tuned to accurately simulate the blade's performance. In conclusion, the implemented grid sensitivity analysis proves that the mesh generated by the Turbo-Grid module is effective and efficient in simulating the blade's behaviour.

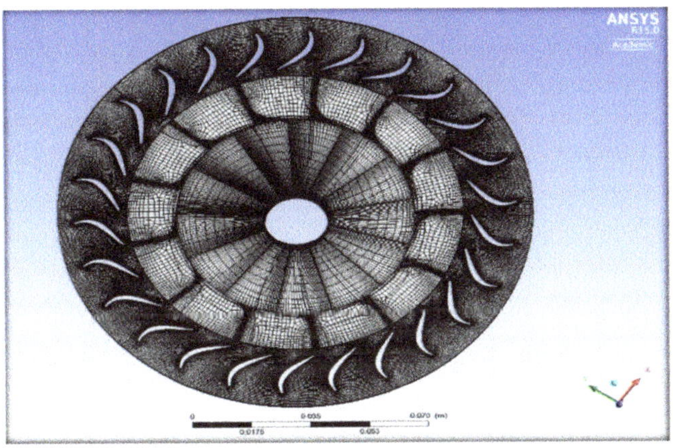

Figure 8. The 3D mesh generations for the stator and the rotor together.

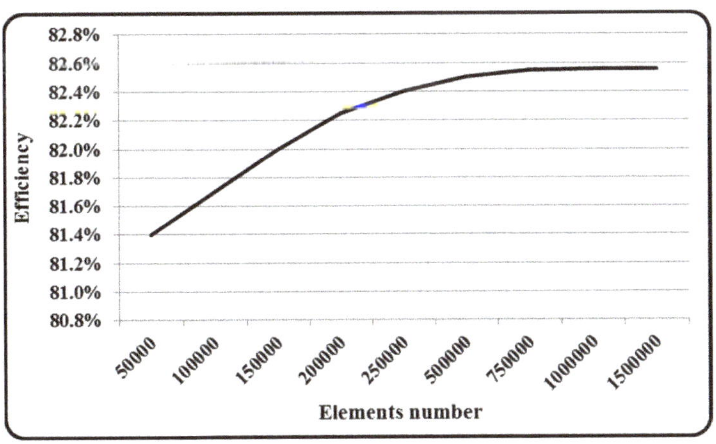

Figure 9. Turbine model efficiency corresponding to element number.

The quality of the solution is one of the most important factors that directly affects both the results accuracy on the one hand and computational costs on the other [27,28]. Using the ANSYS®-CFX solver, an aerodynamic analysis of 3D turbulent viscous flow for the complete stator and rotor SSRT model was achieved. The utilized settings were chosen to be steady-state 3D viscous, single phase and compressible flow as the compressed air was the working fluid. As suggested by the authors of reference [29], the first order upwind advection scheme with topological equation was also selected during the settings as this helps with gaining a high stability in the analysis. Moreover, the stage interface was employed as the analysis and this study focused on the complete unit, stator and rotor and the coupling of shear stress turbulence model; the Navier–Stokes equations were selected. The CFX-Solver theory guide [6,30] recommended preserving the y^+ average value regarding unity. Regarding the boundary conditions, the total temperature, total pressure, flow direction and the rotational speed were also selected and fed to the software as inlet boundary conditions and, at the same time, the static pressure was nominated to represent that of the output. Finally, a rotational and adiabatic wall were picked for the blade's surface and the hub, respectively.

Based on the transport equations of k-ω, as depicted in Equations (17) and (18), the convergence conditions for the velocity and continuity were established at a magnitude of 10^{-5}, whereas the energy equation was selected with an order of 10^{-6}. Upon the fulfilment of the aforementioned criteria, the requisite solutions were attained.

The transport equations for the k-ω model are presented below:

$$\frac{\partial}{\partial t}(\rho k) + \frac{\partial}{\partial x_i}(\rho k u_i) = \frac{\partial}{\partial x_j}\left(\Gamma_k \frac{\partial k}{\partial x_j}\right) + G_k - Y_k + S_K \tag{17}$$

$$\frac{\partial}{\partial t}(\rho \omega) + \frac{\partial}{\partial x_i}(\rho \omega u_i) = \frac{\partial}{\partial x_j}\left(\Gamma_k \frac{\partial \omega}{\partial x_j}\right) + G_\omega - Y_\omega + S_\omega \tag{18}$$

Once the aerodynamic analysis was successfully achieved, the SSRT, including the designed volute, the stator and the rotor, were modelled using the Mechanical Workbench of ANSYS® in order to extract the solid model from their air domains. The second important step was designing the other required turbine parts such as the shroud and the inlet and exit flanges with the assistance of SOLIDWORKS (version 2016) software.

Then, the CFD model and the static-structural model were coupled with the purpose of evaluating the turbine's stress and deformation values. This could be achieved when the aerodynamic pressure values, which were calculated in the CFD model, were transferred through system coupling to the structural model. Also, the temperature value, which the rotor design was initially based on, was coupled to the solid model to assess the turbine thermal analysis using ANSYS/Steady State Thermal section. An adequate 3D solid element distribution for the rotor blades and the hub was selected. In order to accurately complete the structural analysis, some areas of interest needed fine meshing.

The other important issue at this stage was defining the related (Mechanical, Fatigue and Thermal) properties of the new material, RGD 525, and inserting them into the software [31]. Some of those properties were found in the supplier data sheet and others were obtained using some experimental tests, such as the fatigue and the tensile test at various temperature values. Another crucial factor is whether a complete mesh is needed for the whole rigid body or merely the surface contact mesh. This is controllable by stiff body behaviour. Clearly, the decision of a dimensionally decreased model has been made in this investigation to reduce the computational time required. The second essential aspect in mesh analysis pertains to the transition ratio, which can be defined as the governing factor that dictates the rate of expansion of adjacent elements, encompassing a numerical range from 0 to 1. The present investigation has chosen a value of 0.272, as stated by the authors of reference [29]. The appropriate regions of the hub and blade surfaces were subsequently chosen to prevent any displacement as the primary means of support for the structure. Additionally, the pressure side of the rotor blade was selected to apply centrifugal forces, taking into account the angular velocity derived from the initial aerodynamic analysis. It is imperative to acknowledge presently that the temperature of the working fluid, specifically the compressed air, was incorporated into the structural model. Table 2 provides the parameters for the specific structural mesh that was chosen, while Figure 10 depicts the distribution of element density across the SSRT model assembly.

Table 2. Model mesh parameters.

Physics Preference	Mechanical
Sizing:	
Relevance Centre	Medium
Initial Size Seed	Active Assembly
Smoothing	Medium
Transition	Fast
Span Angle Centre	Coarse
Minimum Edge Length	0.0546 mm
Inflation:	
Inflation Option	Smooth Transition
Transition Ratio	0.272
Maximum Layers	5
Growth Rate	1.2
Inflation Algorithm	Pre
Patch Independent Options:	
Topology Checking	Yes
Advanced:	
Shape Checking	Standard Mechanical
Element Mid-side Nodes	Program Controlled
Extra Retries for Assembly	Yes
Rigid Body Behaviour	Dimensionally Reduced

Figure 10. The cross-sectional view of the designed SSRT showing its mesh.

The utilization of time stepping was not implemented in the analysis due to its static nature. However, the coupling setting procedure was enabled to facilitate the activation of the two-way coupling analysis technique. The present analysis utilizes a Computational Fluid Dynamics (CFD) model to evaluate the extraction of aerodynamic loads. This is achieved through the implementation of fluid–solid interfaces, which facilitate the transfer of these loads to the blades. The evaluation of the mechanical loads, deformations, material strains and fatigue properties will be performed following the selection of the requisite material attributes.

The following analysis, which needed to be evaluated, was the modal analysis utilizing FEM through the ANSYS software [30] and using 3D 8-nodes solid elements with linear interpolation [24]. The model had a total of 997,735 elements and 1,995,470 nodes. A cantilevered support type was applied in order to idealize the boundary condition of the model when the blade body was fixed on the rotor hub. In this type of analysis, the actual scales of the deformations and any resulting derived values, like strains and stresses, are meaningless because the solution in a modal analysis consists of a deformed shape that is arbitrarily scaled. So, only the relative magnitudes of such quantities will be measured as meaningful values using this model. Understanding this analysis, and how the structure will vibrate if it is excited, will, in fact, enable the designer to design better structures [24]. This occurs at or near the mode natural frequency when the overall vibration shape of a structure has a tendency to be dominated by the resonance mode shape [30].

4.2. SSRT Multidisciplinary Optimization

Even if the aerodynamic optimization across all stages of the SSRT was successfully accomplished and concluded in the previous chapter, the SSRT must be concurrently analysed from a structural viewpoint using the FEA approach in order to obtain its optimal design. As previously stated, the optimized rotor shape underwent structural analysis, during which several key parameters were parameterized to improve the structural integrity of the rotor. This resulted in reduced stresses, vibrations and deflection. The proposed optimization technique successfully combines computational fluid dynamics (CFD) modelling and finite element analysis (FEA) to achieve optimal results in terms of turbine efficiency, power output and stress and deflection limitations, without any compromise. The input components for this study included the blade thickness along the blade, the solid mass and the hub offset. The output parameters considered in the analysis were the turbine output power and efficiency, as determined through aerodynamic research. Additionally, the equivalent maximum stress and deflections were evaluated through structural analysis, along with the turbine's estimated lifespan. The achievement of the final turbine form poses a significant challenge and necessitates a careful balance between the specified parameters. This can be effectively accomplished through the utilization of multi-objective interdisciplinary optimization techniques [32].

5. Results and Discussion

5.1. Experimental Validation of the CFD Model of SSRT

By manufacturing the Small-Scale Radial Turbine, SSRT, the complete test rig became ready to be used and investigates the model using a varied range of operating conditions, inlet temperature, inlet pressure and mass flow rate of the compressed air as a working fluid. To confirm the methodology employed to enhance SSRT performance, the complete developed turbine was designed, manufactured, as shown in last section, and then tested. With the aim of configuring a big map for the examined SSRT and taking into considerations the material and the lab limitations, the test was established at various operating conditions, inlet temperature, inlet pressure and flow rate values. The employed procedure in that test can be summarized as follows:

1. The compressed air was allowed to flow from its storage tank, at atmospheric temperature, to a pressure regulator, which is used to achieve the required value of the inlet pressure.
2. During its journey to the turbine, the compressed air was heated by passing through the thermal receiver. The required temperature value can be achieved using various levels between the light source and the thermal receiver, as shown in the previous section.
3. By doing so, the compressed air now has both thermal and kinetic energy, and is guided to the expander, representing by the SSRT, in order to extract that energy and convert it into useful energy. As a result, the turbine performance can be examined at different scenarios.
4. By changing the inlet total pressure and inlet mass flow rate, and keeping the inlet temperature of the SSRT fixed, the first one was carried out at different values of compressed air inlet temperature.

5. The second scenario was established by fixing both the inlet mass flow rate and pressure values and changing the SSRT inlet temperature. This was the case at different values of compressed air mass flow rate and inlet pressure.

Also, using the same conditions found in the rig, a three-dimensional numerical analysis, using ANSYS CFX, was simultaneously established to compare the two different types of results.

The graphical representation in Figure 11 illustrates the comprehensive static efficiency and power generation of the SSRT under specific operating conditions, namely with an intake temperature of 40 °C, a maximum rotor speed of approximately 20,000 rpm and various values of Pressure Ratio PR. A comparison is made between the theoretical predictions and the corresponding experimental results. It is important to acknowledge that the highest level of efficiency and power generation was achieved when the intake pressure reached approximately 3 bar and the inlet turbine temperature was maintained at 40 °C, resulting in respective values of 45 percent efficiency and 1.05 kW power output. The data presented in this figure suggest that the calculated efficiency and output power of the SSRT were overestimated, with a maximum value of 16% and less than 15%, respectively, in comparison to the results obtained from the testing process. This is relevant for the following reasons:

- The numerical investigation was conducted under ideal boundary conditions, with a steady-state flow and negligible heat dissipation through the turbine structure. However, the real-world operating conditions differed significantly.
- The measurements in the experimental study encompassed the surface roughness of the SSRT model, whereas the numerical analysis conducted with ANSYS CFX did not account for this aspect.

Another justification pertains to the mechanical losses occurring in the rotor section, which were not taken into consideration in the CFX analysis. This is because the existing model solely accounted for the stator and rotor, neglecting these losses.

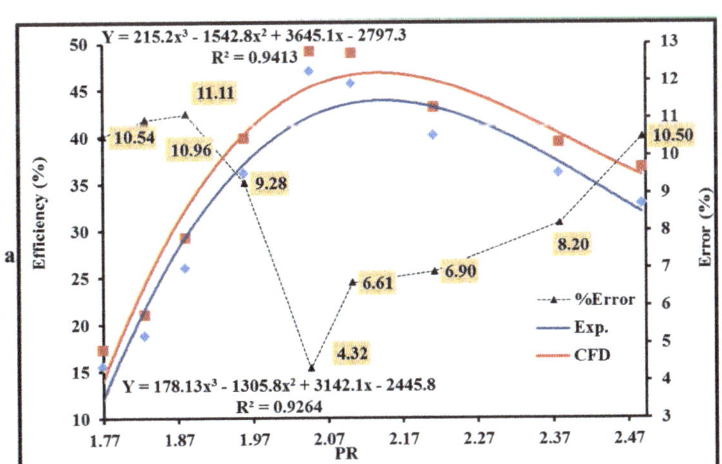

Figure 11. *Cont.*

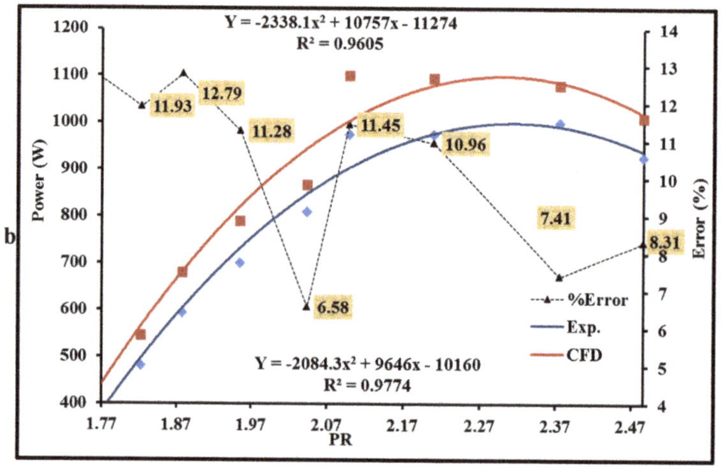

Figure 11. Validation of CFD modelling and experiment for SSRT: (**a**) efficiency and (**b**) developed power at 40 °C.

5.2. Structural Analysis Results

Figure 12 shows the pressure distribution and the velocity stream in a blade-to-blade view for the used blade-to-blade passage for the SSRT. It is clear from the figure how the pressure value decreases through each component of the SSRT, starting from the stator which has the highest value. As for Figure 13b, the velocity contours through the stator and the rotor are presented. From this contour, the maximum value of the compressed air velocity is located at the stator outlet when it is directed toward the rotor inlet, and then decreased gradually when it passes through the rotor to reach its minimum value at the rotor exit.

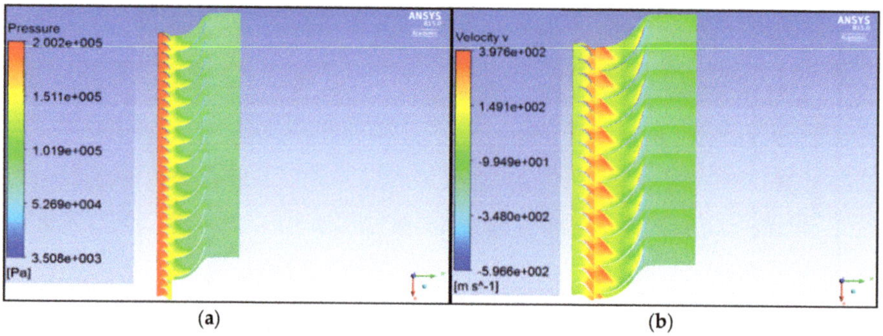

Figure 12. Pressure contours (**a**) and velocity contours (**b**) for SSRT using blade-to-blade vision.

Figure 13 shows the imported pressure and temperature distribution. As shown in the figures, the pressure side of the blade is where the incoming air flow, as an aerodynamic load, is positioned in order to provide the requisite torque. The temperature profile on the blades of the rotor begins at the tip of the blade/maximum temperature value (because the incoming air flow rate has the highest temperature value), and then decreases as the compressed air is distributed on the other blade's surface, particularly at the leading-edge side of the blades.

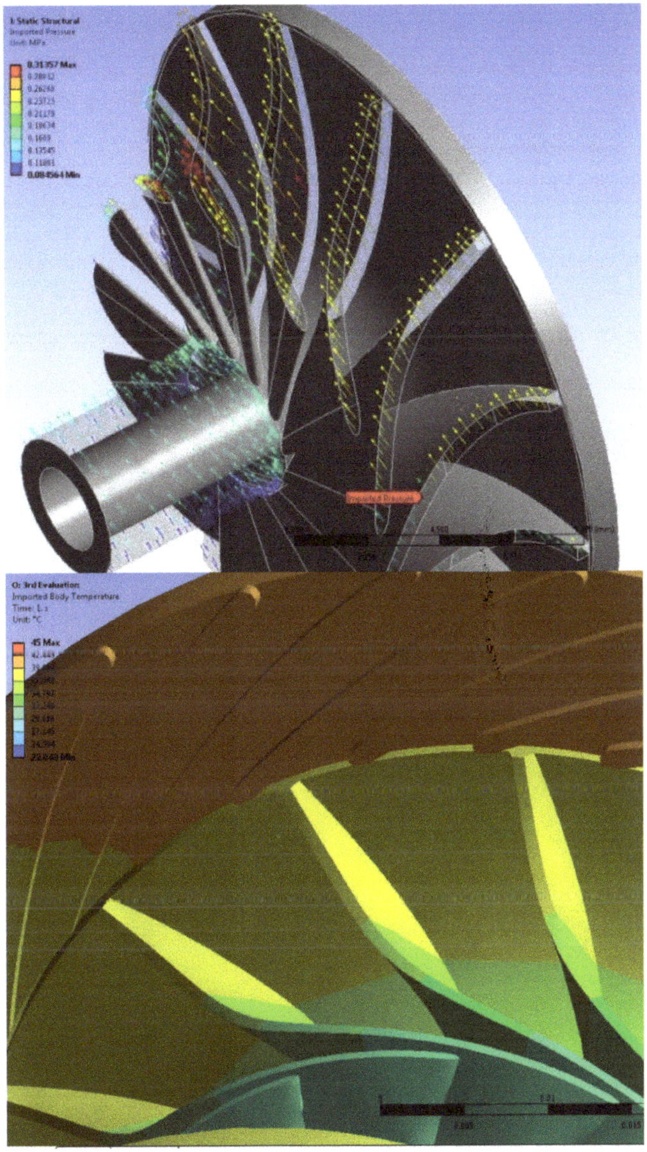

Figure 13. Imported pressure (**top**) and temperature (**bottom**) from the CFD model to the FEA model.

Once these two parts are initiated, the next step is to design the required volute required for distributing the inlet compressed air annually, delivering uniform mass flowrate and guiding it to the stator, ideally without pressure losses [33] and, serially, the rotor.

Turbines are one of the models that deal with high temperature analysis and, as a result, a thermal analysis, which plays a significant role in the expanse of endurance structural stresses, should be deeply evaluated. That which has been mentioned earlier will be even more appreciated once one knows that the material which has been used in manufacturing the turbine model is a plastic one. In this study, it was assumed that the inlet of compressed air was initially heated using a small scale thermal receiver (see more details in references [5,34], as these work are interested in the small-scale solar powered cycle applications).

Based on the experimental analysis carried out, the maximum compressed air temperature reached up to 70 °C; however, for the sake of material ability to withstand values such as these, the range with which this work deals was around 50 °C. This temperature value represents the compressed air inlet temperature in the turbine.

In Figure 14, it can be observed that the air inlet temperature has a significant effect on stress, strain and their distribution on the rotor. Based on the heat energy of the compressed air, the flux amount applied on the turbine model's surfaces was estimated. Also, the convection, radiation and conduction locations were specified, depending on the model configuration, as heat sinks.

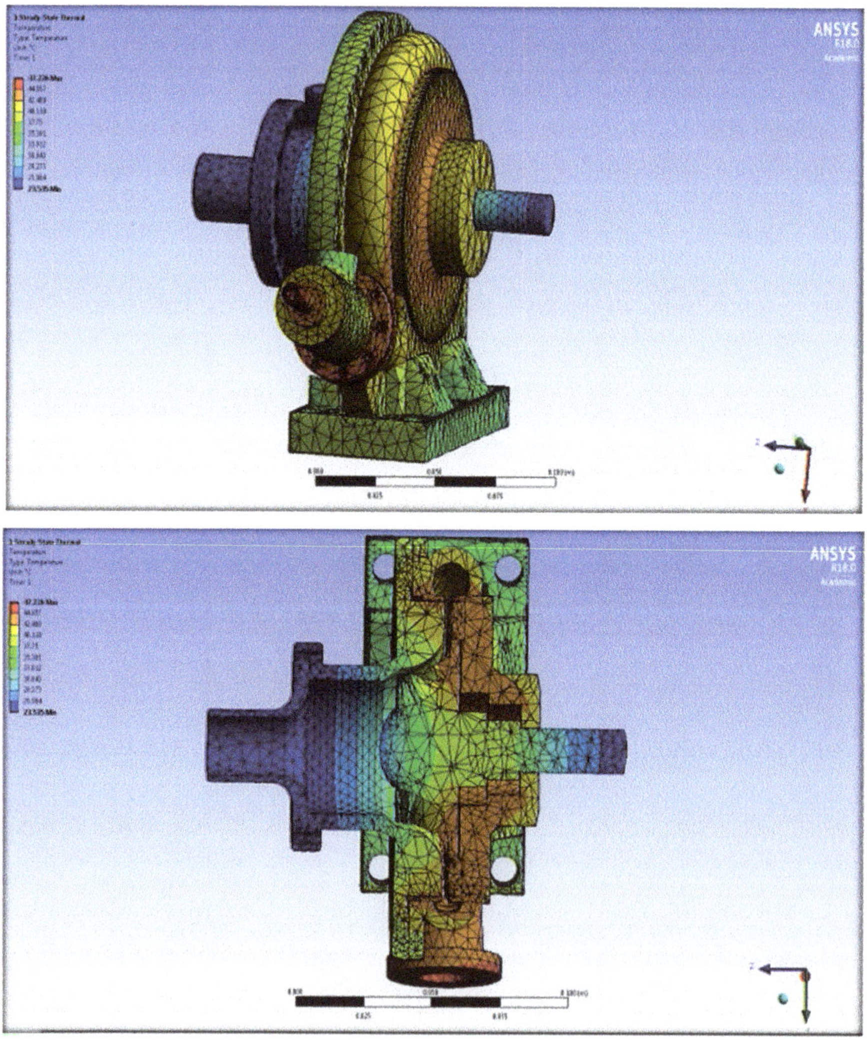

Figure 14. The temperature distribution as a result of heated compressed air supplied through the inlet.

For example, the maximum temperature, around 48 °C, was located at the inlet flange and determined as being the first part that the incoming compressed air faces when entering the turbine model. By contrast, for the parts which are exposed to the atmosphere or those which are located at a further point, their temperature values are lower because of heat loss. Interestingly, the temperature value at the rotor exit is around 23 °C, which is almost the atmospheric temperature as initially defined in the boundary conditions on the assumption that the rotor is extracting most of the compressed air energy, including thermal energy. Here, it is important to highlight that, at this stage, no deformation or stress analyses have yet been established.

The other important consideration in this study is the structural analysis which includes the stress, the deformation and the safety factor values. So, by specifying the inlet and the exit pressure values through aerodynamic analysis, using ANSYS CFX, accompanied by the thermal load along the turbine model, the previously mentioned factors were determined at fixed rotational velocity at this stage. Figure 15 highlights the total deformation that can occur in the SSRT parts. From this figure, it can be noticed that the most influenced part was the rotor, especially the blade tip areas, as it is considered a cantilever beam, while the other rotor's areas were relatively high compared to other parts of the turbine. By contrast, the latter experienced very low deformation because they were fixed, either by another turbine's parts, by bolt connections or to the basement. The other important factor in this analysis is the equivalent stress initiated in the model, which is shown in Figure 16. This stress was concentrated where the blades' roots were located, indicated that this stress was intensified in the previously mentioned area, as well as in the bolt spaces as a result of the applied forces' directions.

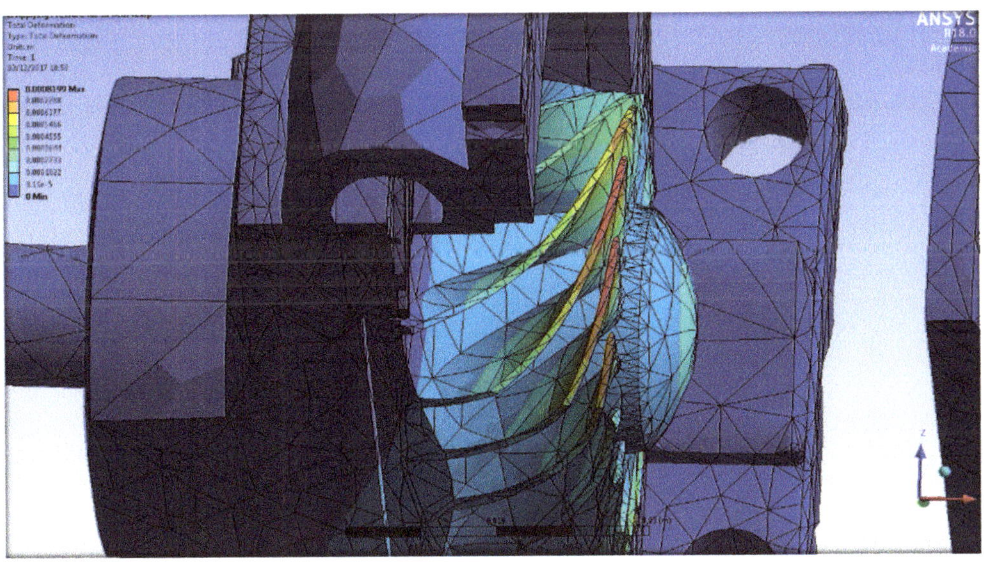

Figure 15. The total deformation distribution of the SSRT.

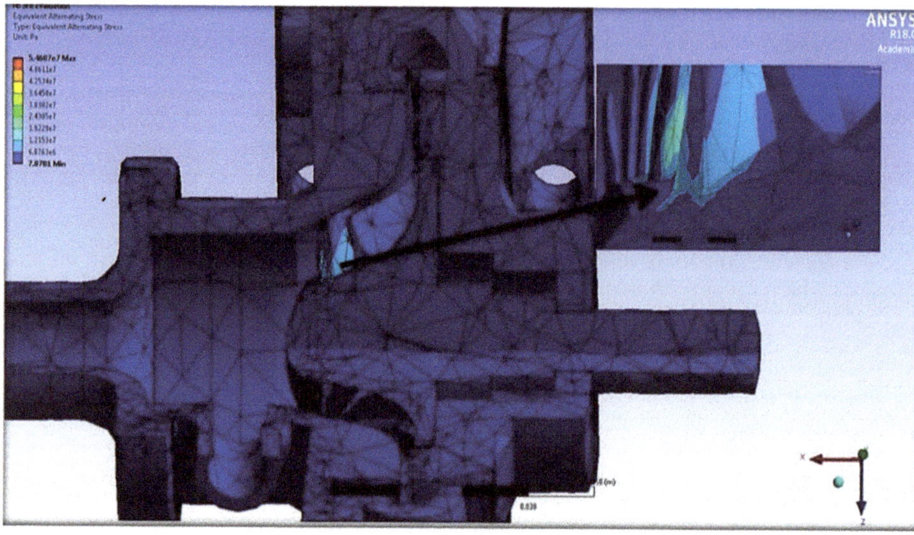

Figure 16. The total equivalent stress distribution of the SSRT.

Figure 17 depicts the comprehensive fatigue assessments conducted on the rotating component of the SSRT, considering the two most extreme input compressed air temperature values of 25 °C and 65 °C. The initial factor assessed pertained to the lifespan, which reveals the notable influence of thermal occurrences on the overall efficacy of the rotor. The design lifespan, initially estimated at approximately 115,000 cycles at a temperature of 25 °C, diminishes to only about one-third of its previous value when subjected to a temperature of 65 °C. In this mode of analysis, the assessment of life at a given moment is conducted under the condition of constant amplitude. The evaluation is based on whether the minimum alternating stress, as indicated by the S–N curve, exceeds the corresponding alternating stress. In this study, the investigation also involved the calculation of fatigue damage to the rotor over its operational lifespan. The findings of this analysis confirm that the blade–hub connection area is the most vulnerable section of the rotor, irrespective of temperature conditions. The highest level of damage observed at the temperature of 25 °C was approximately 56,908 while, at a temperature of 65 °C, it reached approximately 7,350,000. This indicates that the observed damage can be attributed solely to the 40 °C difference in the temperature of the working fluid. In relation to the biaxiality factor, which can be defined as the ratio between the highest principal stress and the lowest principal stress, with the exclusion of the main stress when it is close to zero, the maximum biaxiality was observed to be 0.95 at a temperature of 25 °C and 0.99 at a temperature of 65 °C. Furthermore, the pertinent contour plot indicates that the highest degree of biaxiality was concentrated at the point of intersection between the X and Z axes, as observed at both tested temperatures (25 °C and 65 °C). The compressed air flow is perceived as entering the SSRT through the radial X axis and exiting through the axial Z axis, as depicted in the corresponding figure. The investigation of fatigue also considered the comparable alternating stress, which is demonstrated in the fourth figure, as another important aspect. This factor can be utilized for the identification of the type of fatigue S–N curve. In the context of fatigue analysis, the evaluation of fatigue loading type, mean stress effects and multiaxial effects is a crucial undertaking. Hence, the alternating stress that is equivalent can be considered as the ultimate value computed prior to the computation of fatigue life. The main advantage of this component is its ability to incorporate all fatigue-related calculations regardless of the specific characteristics of the fatigue material. This sets it apart from tensile strength, which is influenced by material properties and relies on other mean stress theories. The maximum values of this component have been observed to reach approximately 54.7 MPa and 57.9 MPa at the two tested temperatures of compressed air, specifically 25 °C and 65 °C, respectively. Finally, the fatigue safety factor, which differs to some extent from the factor estimated under static load conditions due to the dynamic nature of load variation over time, exhibits certain distinctions from the factor calculated under static load conditions. Hence, this particular element is associated with the occurrence of fatigue failure within a designated design lifespan. At a temperature of 25 °C, the minimum value observed was approximately 3.9, which subsequently decreased to below 0.1 as the compressed air progressed to a temperature of 65 °C. However, it is noteworthy that the safety factor of the rotor's other areas exhibited significant variation, ranging from 55 at 65 °C to 3 at 25 °C. In contrast, the highest safety factor was approximately twice as high as that of the majority of these areas.

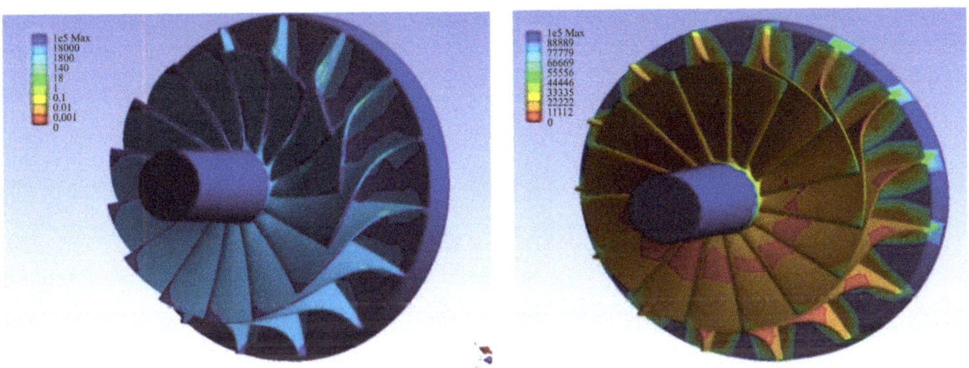

Figure 17. An example of fatigue life expectation in the rotor at two different values of inlet air temperature; 25 °C (**Left-Hand side**) and (**Right-Hand Side**) 65 °C.

Figure 18a illustrates the maximum deformation observed at the blade tip in response to an augmentation in the rotational velocity of the rotor. The aforementioned element demonstrated a notable response to the escalation in rotor velocity. At a rotational speed of 60,000 rpm, the most significant deformations were observed at approximately 0.401, 0.592 and 0.874 mm for fluid temperatures of 25, 45 and 65 °C, respectively. The recorded values for the measurements were approximately 0.52 mm, 0.63 mm and 0.87 mm when the rotational speed was set at 80,000 rpm. The relatively small displacement values observed can be attributed to the turbine's small dimensions. In analysing the impact of fluid temperature on rotor displacement, it becomes apparent that this particular factor had a negligible influence. This phenomenon could potentially be attributed to the material's capacity to maintain stable temperatures within the specified range. The fourth parameter examined in this research pertains to the fatigue experienced in the rotating component of the analysed SSRT. The relationship between the rotational speed of the rotor and the three input air temperatures is depicted in Figure 18B. The presented graph illustrates the correlation between rotational speed, temperature and the expected fatigue life of the rotor. As an illustration, when the rotor rotational speed reached 60,000 rpm, the number of cycles within the endurance limit was approximately 140,000, 88,000 and 71,000 at compressed air temperatures of 25, 45 and 65 °C, respectively. At a rotational speed of 80,000 rpm, the corresponding values at temperatures of 25, 45 and 65 °C were recorded as 97,000, 57,000, and 50,000 cycles, respectively. It is imperative to acknowledge that the statistics provided herein represent the minimum quantity of cycles, and that the location potentially susceptible to fatigue life is depicted in the preceding images. The fatigue life values observed at a rotational speed of 70,000 rpm were found to be approximately equidistant from the values observed at the other two temperatures. The values for compressed inlet temperatures of 25, 45 and 65 °C are approximately 109,000, 71,000 and 59,000, respectively. Figure 18C illustrates the comprehensive extent of damage observed in the rotor section of the investigated SSRT, with respect to variations in both the temperature of the working fluid and the rotational speed of the rotor. Significant damage was primarily observed when the temperature and rotating speed reached their highest levels. The phenomenon under consideration can be elucidated through an examination of the interplay between thermal and mechanical stresses and their impact on the structure of the rotor. The maximum damage values recorded at a rotational speed of 60,000 rpm were approximately 51,400, 841,000 and 7,140,000 for working fluid temperatures of 25, 45 and 65 °C, respectively. In contrast, during the highest examined rotational speed of 80,000 rpm, the corresponding damage values were approximately 260,000, 111,000 and 4,390,000 for working fluid temperatures of 25, 45 and 65 °C, respectively. Furthermore, when the rotor rotational speed reached 70,000 revolutions per minute, the observed damage values were approximately 125,000, 952,000 and 4,100,000 for working fluid temperatures of 25, 45 and 65 °C, correspondingly.

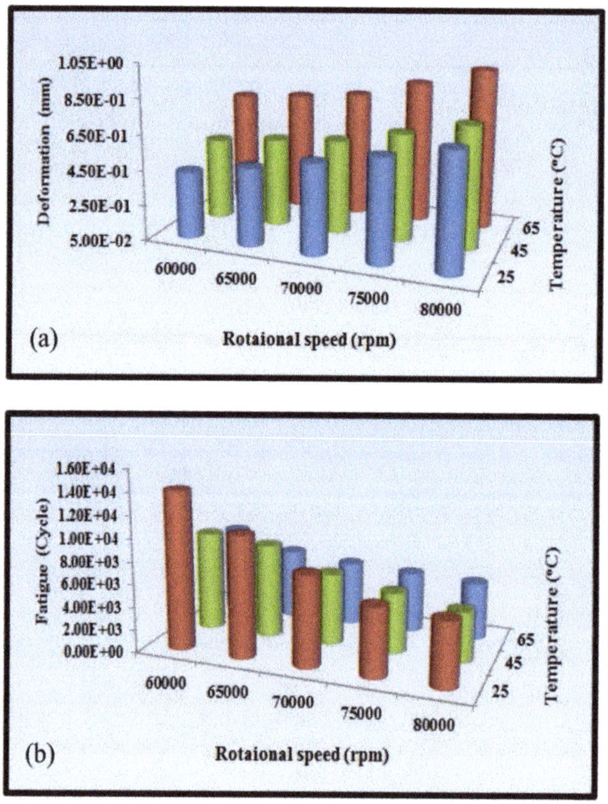

Figure 18. *Cont.*

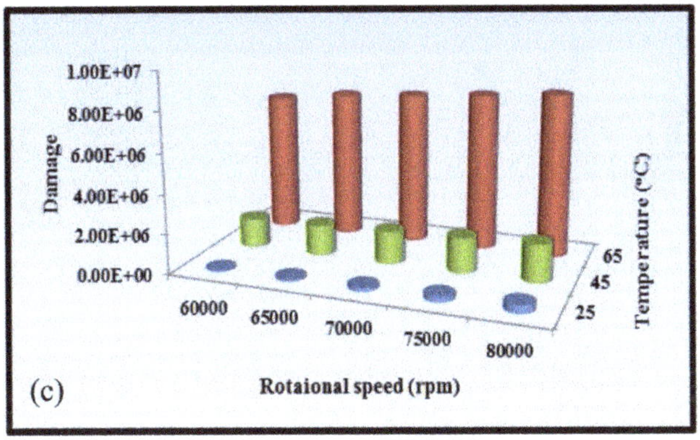

Figure 18. The effect of both inlet air temperature and the rotor rotational speed on: (**a**) the maximum deformation, (**b**) the rotor fatigue life and (**c**) the overall damage in the rotor's body. Different coloured columns represent variables at different temperatures.

5.3. Multidisciplinary Optimization Results

The final step in this work is to find the optimized turbine shape that is able to obtain the highest values of both the turbine efficiency and power output and, at the same time, able to withstand the stresses, deflections and fatigue. So, as previously mentioned, the most important parameters, which directly affected the above-mentioned analyses, were considered in order to structurally enhance the rotor performance and, at the same time, conserve its aerodynamic behaviour, as it achieved the best results in this regards. Figure 19 depicts the chart of parallel parameters for the SSRT using multidisciplinary optimization, while Figure 20 depicts the chart of design points with some of the selected parameters for the SSRT utilizing multidisciplinary optimization. Figure 21 illustrates the sensitivity of output parameters to each of the selected input factors in order to identify the most influential parameters. The graph illustrates that the rotational speed of the rotor is a significant determinant of the maximum equivalent stress and subsequent fatigue of the turbine rotor. The rotor stagger angle emerged as a noteworthy input parameter, exerting a substantial influence on both the efficiency of the turbine and the level of stress experienced by the rotor. In relation to aerodynamic performance, it can be observed that the thickness of the blade exerts a more significant influence on the stress value compared to other outlet factors.

The proposed correlation between the rotor rotational speed and the minimum life is presented in Figure 22. This shows the importance of having relatively low values of rotational speed in order to achieve the highest level of rotor life, as a result of having lower equivalent stress value. The linear correlation is imported as $Y = -0.013072X + 374.84$ and the $R^2 = 99.165\%$. However, the quadratic trend line is $Y = -3.1132 \times 10^{-6}X^2 + 0.027424X + 243.59$ and the $R^2 = 99.773\%$ when the Y and X are the minimum required life and the rotor rotational speed, respectively, where the proposed correlation is relatively high. The final step in this section is to find the rotor blade shape which has the lowest stress and deflection values and, at the same time, is able to withstand rotation for the longest time with as high a turbine efficiency and output power as possible, preserving the aerodynamic performance. The rotor exhibited a notable improvement in terms of the two most essential structural analysis criteria, the maximum equivalent stress and the fatigue life, with the first output parameter being lowered in order to achieve a greater number of cycles. Nonetheless, the consequence was a drop in total turbine efficiency of around 2%.

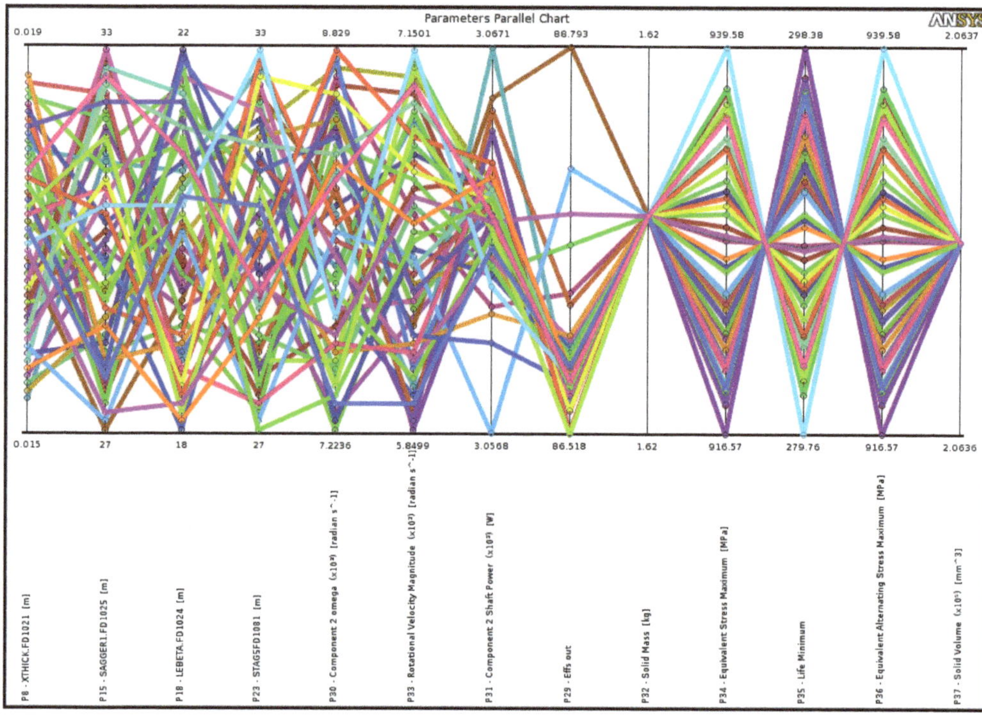

Figure 19. Parameters parallel chart for the SSRT using the multidisciplinary optimization. Each coloured-line represents different design parameter.

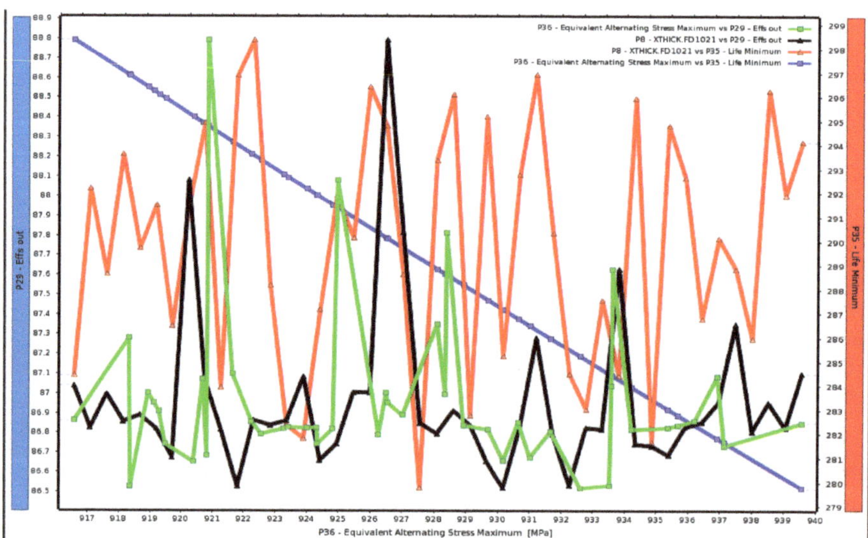

Figure 20. Design points vs. chosen parameters chart for the SSRT using multidisciplinary optimization.

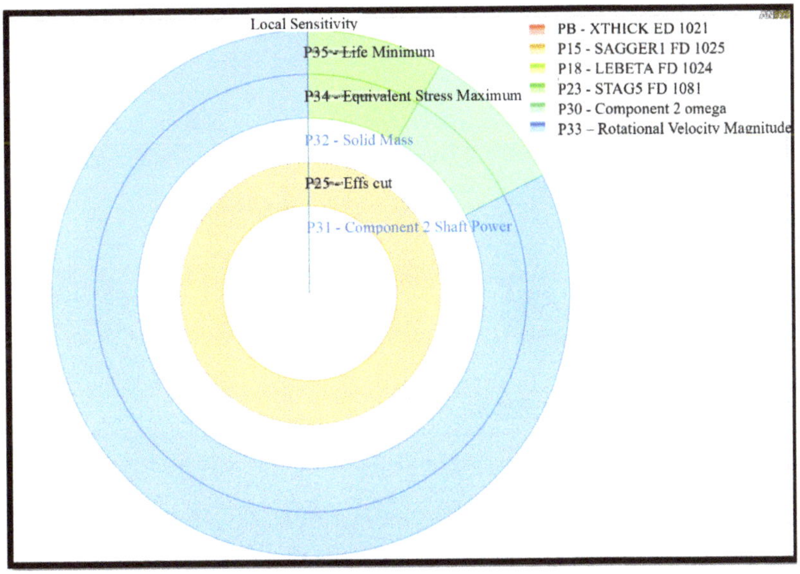

Figure 21. The local sensitivity of the outlet parameters with respect to the inlet parameters.

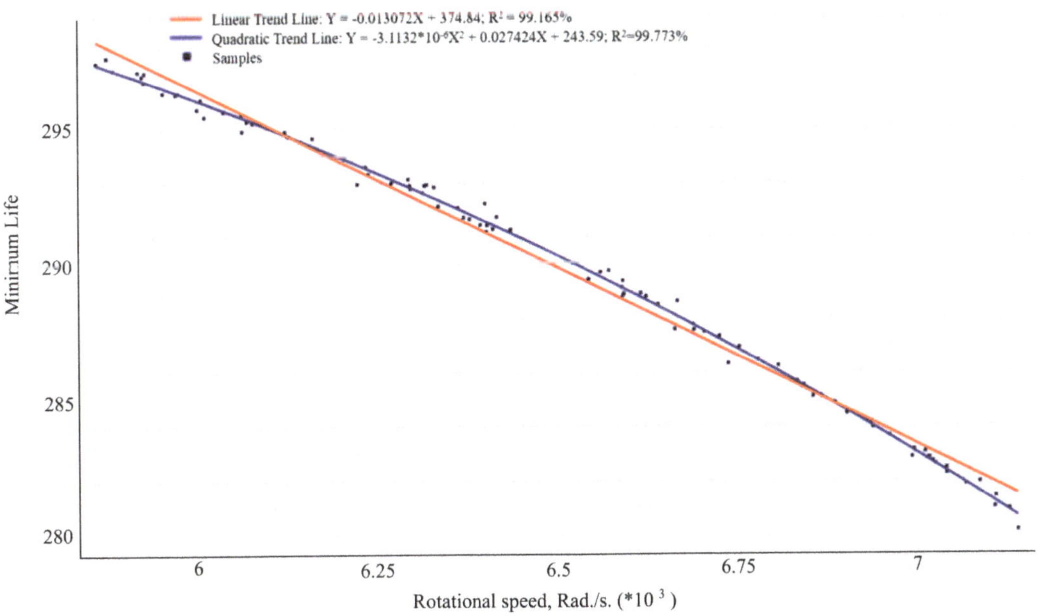

Figure 22. The achieved correlation between the rotor speed and the minimum life.

Figures 23–25 show the values and distributions of equivalent stress, fatigue life and displacement, respectively, for interdisciplinary optimization. Figure 26 details the trade-off between the maximum equivalent stress, the fatigue life, the efficiency and the output power for the SSRT.

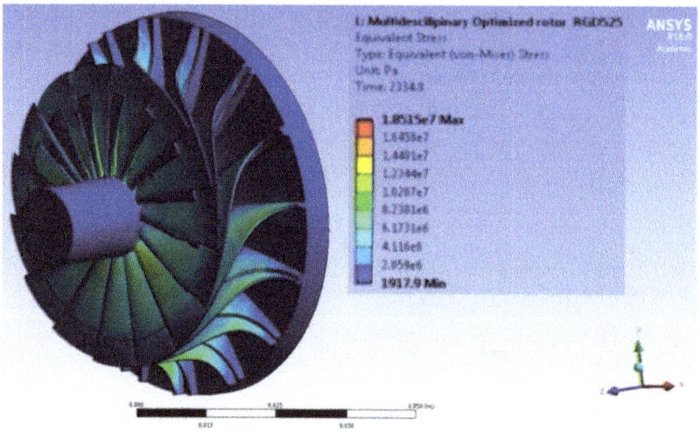

Figure 23. Maximum equivalent stress value for the multidisciplinary optimized rotor of the SSRD.

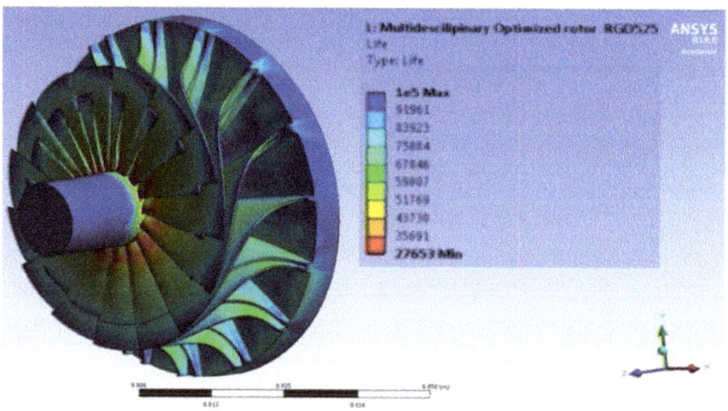

Figure 24. The fatigue life for the multidisciplinary optimized rotor of the SSRD.

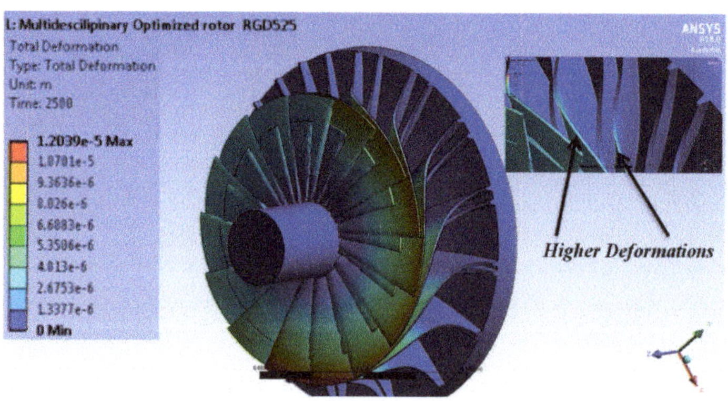

Figure 25. The total deformation for the multidisciplinary optimized rotor of the SSRT.

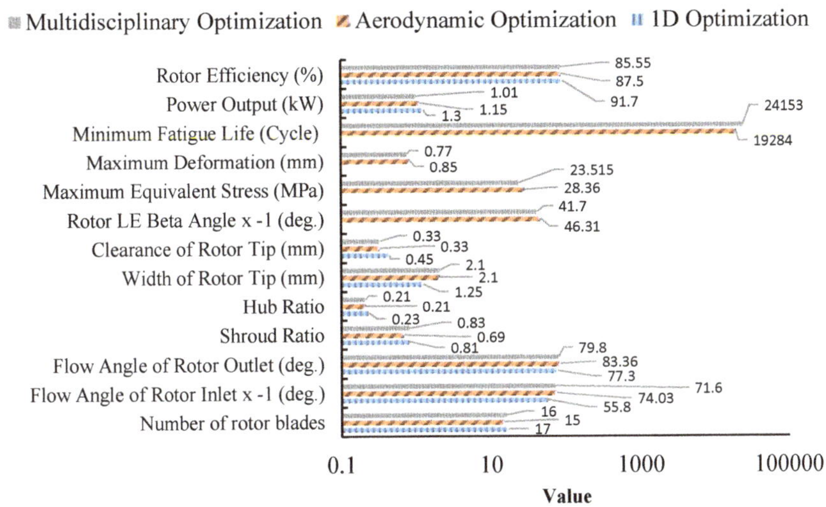

Figure 26. The chosen parameters compromission.

6. Conclusions

The present study extensively examines the impact of several significant boundary conditions, namely the inlet temperature of the compressed air, the rotational speed of the turbine's rotor and the working pressure ratio, on the performance of the Small-Scale Radial Turbine. This investigation was conducted using 3D analysis and focused on its application in small-scale solar-powered Brayton cycle systems. The analysis builds upon prior research conducted by the authors in order to improve the aerodynamics of the turbine under study. The most significant consequences of this research can be summarized as follows:

1. It was observed that the rotating speed of the rotor has a considerable influence on the amount of stress and displacement, with a maximum increase of 69% in stress and 59% in deformation seen at the highest examined rotational speed of 65 °C.
2. The temperature of the fluid must also be carefully examined as a significant component. It was shown that reducing the temperature of compressed air to 25 °C reduced the aforementioned increases to about 27% and 7%, respectively.
3. In terms of stress region, the stress concentration was mostly in the region between the hub and the blades; thus, this area must be strengthened to survive high stress intensity in this position.
4. The region of the tip shroud in the rotor had greater deflection values at 21% of the blade tip width. Consequently, this distance between the blade tip and the shroud, as well as in between the blades, should be sufficient to accommodate the deflection.
5. According to the fatigue study, the increased input temperature of the fluid and compressed air resulted in an 84% decrease in rotor fatigue life, particularly at higher rotational speeds. The region where the rotor's blades link to the hub had the lowest fatigue life. This location sustained the most damage among the other rotor bodies.
6. A structural analysis needs to be carefully and simultaneously considered during aerodynamic analysis in order to sustain what has been achieved in terms of aerodynamic analysis. Consequently, multi-objective optimization should also include some structural parameters and some of the objective function is needed in order to cover the stresses in other structural analyses.

Author Contributions: Conceptualization, H.H.; Methodology, A.M.D.; Software, A.M.D.; Validation, A.M.D. and M.A.B.; Investigation, A.M.D. and A.A.H.; Resources, S.S.; Data curation, A.K.; Writing—original draft, A.M.D.; Writing—review & editing, A.A.H.; Visualization, T.B.; Supervision, S.M. and R.A.-D. All authors have read and agreed to the published version of the manuscript.

Funding: This research received no external funding.

Data Availability Statement: Data is unavailable for public access.

Acknowledgments: The authors thank the Higher Committee of Developing Education in Iraq HCED for funding the work. They also thank the University of Mosul, the University of Birmingham and Teesside University for the facilities provided for the present research study.

Conflicts of Interest: The authors declare no conflict of interest.

Nomenclature

D	Stiffness matrix
ε	Mechanical strain
s	Stress
E	Modulus of elasticity
α	Coefficient of thermal expansion
ν_p	Poisson's ratio
T	Temperature
m	Mass
r	Radius of rotation (distance between the rotor center of gravity and its rotation centre)
ω_s	Rotor rotational speed
F_{cf}	Centrifugal force
ρ	Material density
A	The blade cross sectional area
z	The blade thickness
A_t	Cross sectional area of blade at the tip
A_r	Cross sectional area of blade at the root
lb	The blade length
ω_i	ith natural circular frequency (radians per unit time)
t	Time
f_i	ith natural frequency (cycles per unit time)
N_f	The number of cycles to failure
$\sigma_f{}'$	Fatigue strength parameter
b	Fatigue life exponent
σ_u	The ultimate strength of the material
σ_a	The alternating stress
σ_m	The mean stress
σe	The endurance stress (endurance limit for completely reversed loading)
R	Stress ratio
n_i	ith fatigue cycle
D_f	Material damage due to fatigue
k	Turbulent flow kinetic energy
ω	Specific dissipation rate
Y_k	Dissipation of k
Y_ω	Dissipation of ω
S	User-defined source term
G_k	The generation of k due to mean velocity gradients
$G\omega$	The generation of ω
Γ_k	The effective diffusivity of k
Γ_ω	The effective diffusivity of ω

References

1. Smadi, Y.; Alsood, E.; Aljaradin, M. A Solar Disinfection Water Treatment System for Rural Areas/Jordan. *Al-Kitab J. Pure Sci.* **2023**, *5*, 55–67. [CrossRef]
2. Mahmood Ibrahim, A.; Ibraheem, R.R.; Bakr Weli, R. Energy Saving in Batteries Using the Photovoltaic System. *Al-Kitab J. Pure Sci.* **2023**, *4*, 78–94. [CrossRef]
3. Edirisinghe, D.S.; Yang, H.S.; Gunawardane, S.D.G.S.P.; Alkhabbaz, A.; Tongphong, W.; Yoon, M.; Lee, Y.H. Numerical and Experimental Investigation on Water Vortex Power Plant to Recover the Energy from Industrial Wastewater. *Renew. Energy* **2023**, *204*, 617–634. [CrossRef]
4. Alkhabbaz, A.; Yang, H.S.; Weerakoon, A.H.S.; Lee, Y.H. A Novel Linearization Approach of Chord and Twist Angle Distribution for 10 KW Horizontal Axis Wind Turbine. *Renew. Energy* **2021**, *178*, 1398–1420. [CrossRef]
5. Daabo, A.M.; Al-Mola, Y.S.; Al-Rawy, A.Y.; Lattimore, T. State of the Art Single-Objective Optimization of Small Scale Cylindrical Cavity Receiver. *Sustain. Energy Technol. Assess.* **2019**, *35*, 278–290. [CrossRef]
6. Daabo, A.M.; Bellos, E.; Pavlovic, S.; Bashir, M.A.; Mahmoud, S.; Al-Dadah, R.K. Characterization of a Micro Thermal Cavity Receiver–Experimental and Analytical Investigation. *Therm. Sci. Eng. Prog.* **2020**, *18*, 100554. [CrossRef]
7. Zhang, R.; Xin, Z.; Huang, G.; Yan, B.; Zhou, X.; Deng, X. Characteristics and Modelling of Wake for Aligned Multiple Turbines Based on Numerical Simulation. *J. Wind Eng. Ind. Aerodyn.* **2022**, *228*, 105097. [CrossRef]
8. Micallef, D.; Ferreira, C.; Herráez, I.; Höning, L.; Yu, W.; Capdevila, H. Assessment of Actuator Disc Models in Predicting Radial Flow and Wake Expansion. *J. Wind Eng. Ind. Aerodyn.* **2020**, *207*, 104396. [CrossRef]
9. Collings, P.; Yu, Z. Modelling and Analysis of a Small-Scale Organic Rankine Cycle System with a Scroll Expander. *Lect. Notes Eng. Comput. Sci.* **2014**, *2*, 1472–1477.
10. Daabo, A.M.; Mahmoud, S.; Al-Dadah, R.K.; Al Jubori, A.M.; Bhar Ennil, A. Numerical Analysis of Small Scale Axial and Radial Turbines for Solar Powered Brayton Cycle Application. *Appl. Therm. Eng.* **2017**, *120*, 672–693. [CrossRef]
11. Gomes, R.P.F.; Henriques, J.C.C.; Gato, L.M.C.; Falcão, A.F.O. Multi-Point Aerodynamic Optimization of the Rotor Blade Sections of an Axial-Flow Impulse Air Turbine for Wave Energy Conversion. *Energy* **2012**, *45*, 570–580. [CrossRef]
12. Shanechi, M.M.; Odabaee, M.; Hooman, K. Optimisation of a High Pressure Ratio Radial-Inflow Turbine: Coupled CFD-FE Analysis. *Proc. ASME Turbo Expo* **2015**, *2C*. [CrossRef]
13. Barsi, D.; Perrone, A.; Ratto, L.; Simoni, D.; Zunino, P. Radial Inflow Turbine Design through Multi-Disciplinary Optimisation Technique. *Proc. ASME Turbo Expo* **2015**, *8*, V008T23A009. [CrossRef]
14. Wang, H.L.; Xi, G. Effect of Thermal and Mechanical Loads on the Centrifugal Impeller Deformation and Its Structure Optimization. *Proc. ASME Turbo Expo* **2009**, *5*, 75–84. [CrossRef]
15. Fu, L.; Shi, Y.; Deng, Q.; Li, H.; Feng, Z. Integrated Optimization Design for a Radial Turbine Wheel of a 100 KW-Class Microturbine. *J. Eng. Gas Turbines Power* **2012**, *134*, 012301. [CrossRef]
16. Alawadhi, K.; Alhouli, Y.; Ashour, A.; Alfalah, A. Design and Optimization of a Radial Turbine to Be Used in a Rankine Cycle Operating with an OTEC System. *J. Mar. Sci. Eng.* **2020**, *8*, 855. [CrossRef]
17. Feng, Z.; Deng, Q.; Li, J. Aerothermodynamic Design and Numerical Simulation of Radial Inflow Turbine Impeller for a 100kW Microturbine. *Proc. ASME Turbo Expo* **2005**, *1*, 873–880. [CrossRef]
18. Causier, A.; Carret, G.; Boutin, C.; Berthelot, T.; Berthault, P. 3D-Printed System Optimizing Dissolution of Hyperpolarized Gaseous Species for Micro-Sized NMR. *Lab A Chip* **2015**, *15*, 2049–2054. [CrossRef]
19. O'Neill, P.; Jolivet, L.; Kent, N.J.; Brabazon, D. Physical Integrity of 3D Printed Parts for Use as Embossing Tools. *Adv. Mater. Process. Technol.* **2017**, *3*, 308–317. [CrossRef]
20. *ANSYS CFX-Solver Theory Guide*; ANSYS, Inc.: Canonsburg, PE, USA, 2009; Volume 15317, pp. 724–746.
21. Singh, M. *Steam Turbines-Design, Application, and Re-Rating*; Mcgraw-hill Education-Europe: Sydney, Australia, 2008.
22. Poblano-Salas, C.A.; Barceinas-Sanchez, J.D.O.; Sanchez-Jimenez, J.C. Failure Analysis of an AISI 410 Stainless Steel Airfoil in a Steam Turbine. *Eng. Fail. Anal.* **2011**, *18*, 68–74. [CrossRef]
23. Dowling, N.; Dowling, N. *Mechanical Behavior of Materials: Engineering Methods for Deformation*; Pearson: London, UK, 2012; p. 59. Available online: https://www.amazon.co.uk/Mechanical-Behavior-Materials-Norman-Dowling/dp/0131395068 (accessed on 28 June 2023).
24. Boentgen, R.R.; Clickner, S.J. Experimental Modal Analysis. *CSI Reliab. Week* **1983**, *1*, 581–584.
25. Eftekhari, M.; Fatemi, A. Creep-Fatigue Interaction and Thermo-Mechanical Fatigue Behaviors of Thermoplastics and Their Composites. *Int. J. Fatigue* **2016**, *91*, 136–148. [CrossRef]
26. Mura, A.; Ricci, A.; Canavese, G. Investigation of Fatigue Behavior of ABS and PC-ABS Polymers at Different Temperatures. *Materials* **2018**, *11*, 1818. [CrossRef] [PubMed]
27. Mohammed, A.A.; Dawood, A.S. Mixed Convection Heat Transfer in a Ventilated Enclosure with and without a Saturated Porous Medium. *J. Porous Media* **2016**, *19*, 347–366. [CrossRef]
28. Alkhabbaz, A.; Yang, H.S.; Tongphong, W.; Lee, Y.H. Impact of Compact Diffuser Shroud on Wind Turbine Aerodynamic Performance: CFD and Experimental Investigations. *Int. J. Mech. Sci.* **2022**, *216*, 106978. [CrossRef]
29. Djodikusumo, I.; Diasta, I.N.; Sanjaya Awaluddin, I. Geometric Modeling of a Propeller Turbine Runner Using ANSYS BladeGen, Meshing Using ANSYS TurboGrid and Fluid Dynamic Simulation Using ANSYS Fluent. *Appl. Mech. Mater.* **2016**, *842*, 164–177. [CrossRef]

30. Ansys, Inc. *ANSYS CFX-Solver Modeling Guide V14.0. ANSYS CFX V14.0*; Ansys, Inc.: Canonsburg, PE, USA, 2011.
31. Yang, H.S.; Alkhabbaz, A.; Edirisinghe, D.S.; Tongphong, W.; Lee, Y.H. FOWT Stability Study According to Number of Columns Considering Amount of Materials Used. *Energies* **2022**, *15*, 1653. [CrossRef]
32. Korte, J.J.; Salas, A.O.; Dunn, H.J.; Alexandrov, N.M.; Follett, W.W.; Orient, G.E.; Hadid, A.H. Multidisciplinary Approach to Linear Aerospike Nozzle Design. *J. Propuls. Power* **2001**, *17*, 93–98. [CrossRef]
33. Moustapha, H.; Zelesky, M.; Baines, N.; Japikse, D. *Axial and Radial Turbines*; Concepts NREC: White River Junction, VT, USA, 2003; p. 10. Available online: https://www.amazon.co.uk/Axial-Radial-Turbines-Haney-Moustapha/dp/0933283121 (accessed on 28 June 2023).
34. Daabo, A.M.; Hammo, K.E.; Mohammed, O.A.; Hassan, A.A.; Lattimore, T. Performance Investigation and Design Optimization of Micro Scale Compressed Air Axial Turbine for Domestic Solar Powered Brayton Cycle. *Sustain. Energy Technol. Assess.* **2020**, *37*, 100583. [CrossRef]

machines

MDPI

Article

Interdependence of Technical and Technological Parameters in Polymer Ultrasonic Welding

Dubravko Rogale [1], Siniša Fajt [2], Snježana Firšt Rogale [1,*] and Željko Knezić [1]

[1] Faculty of Textile Technology, University of Zagreb, 10000 Zagreb, Croatia
[2] Faculty of Electrical Engineering and Computing, University of Zagreb, 10000 Zagreb, Croatia
* Correspondence: sfrogale@ttf.unizg.hr

Abstract: The welding of foils, textiles, and textile composites made of thermoplastic polymer materials using machines with an ultrasonic rotary sonotrode is a high-tech welding technique. Many authors have dealt with only a few parameters in earlier papers, mainly mentioning the speed, i.e., the welding time, and the power of the ultrasonic generator. In this paper, the acoustic model of ultrasonic welding is defined. Based on the model, a group of 44 different parameters important for ultrasonic welding of polymer materials has been summarised, namely 12 parameters of the polymer material, 11 general acoustic and electroacoustic parameters, and 21 technical parameters depending on the ultrasonic machine. Based on this, a comprehensive mathematical derivation was carried out, linking parameter groups with other findings from acoustics, thermodynamics of polymers, and technical and technological parameters of welding polymer materials. The most important parameters are the power of the ultrasonic generator and the welding time, which in practice are adjusted to produce a solid weld. The method of measuring the amplitude of the sonotrode using a photonic sensor is presented in this paper. For 42 groups of welds done at various welding speeds and ultrasonic generator powers, the breaking forces of ultrasonic welds were measured. There are illustrations of power dependence and breaking forces. The accuracy of the mathematical model was confirmed by comparison with the calculation results based on the findings of these measurements.

Keywords: ultrasonic welding; rotary sonotrode; process machine parameters; material parameters; acoustic parameters; mathematical model

1. Introduction

Due to a number of beneficial characteristics, the ultrasonic welding of thermoplastic polymer materials is growing in popularity. It is distinguished by its comparatively low energy usage per welded joint, welding speed, high weld breaking force, simplicity of use, and environmental friendliness, as there are no adverse effects on the health of the employees or dangerous by-products for the environment. Airtight and watertight ultrasonic welds can also be permeable. Since the tightness of the welds is a crucial quality, the manufacturing of protective clothing, disposable hospital gowns, face masks, shoe covers, filters, bags, diapers, sails, heat-insulating air chambers, aerospace, automotive, marine, transportation, sports, and many other applications [1,2] use the ultrasonic technique of long welds in a continuous mode most frequently. The potential uses for this relatively new method are growing every day. The power of the ultrasonic generator for the ultrasonic sonotrode, the compression force of the sonotrode on the material, and the welding time are the three parameters that are most frequently cited in articles discussing the welding technique of polymer materials. This, however, does not provide a comprehensive breakdown of all the variables that come into play while welding using the modern approach indicated.

Under welding parameters, I. Jones [2] mentions amplitude, power welding mode (including hold time in plunge welding), pressure and rotary sonotrode or anvil gap (continuous welding).

In the optimization of ultrasonic welding process parameters C.-C., Kuo et al. [3] uses a flow diagram in which he specifies the parameters of amplitude, weld pressure, and hold time. When analyzing the effect of the factors on ultrasonic seam tensile properties of nonwoven fabrics M. Kayar et al. [4] specify the parameters of the frequency of ultrasonic vibrations, the compression force of the sonotrode and the welding speed. A. Gomer et al. [5], in the research on the fabrication of fiber reinforced plastics by ultrasonic welding, mention the frequency, power, compression force and welding time as well as the amplitude of sonotrode vibrations. In his research on polymer adhesion by ultrasonic welding E. Sancaktr [6] specifies the parameters of frequency, power, amplitude and welding speed. In their research on mechanisms of ultrasonic welding of textile materials W. Shi and T. Little [7] specify speed, compression force, welding time, amplitude and power as process parameters. In their research on ultrasonic welding of all-polypropylene composite, Z. Kiss et al. [8] mention welding time, amplitude, frequency and compression force. In their research on effects of different roller profiles on the microstructure and peel strength of the ultrasonic welded joints of nonwoven fabrics T.-h. Nguyen et al. [9] list frequency, power, amplitude and welding speed as parameters. In their research related to the development of textile-based transmission lines using conductive yarns and ultrasonic welding technology for e-textile applications, O. Atalay et al. [10] mention only strength and compression force as process parameters. Based on a literature review of ultrasonic welding of thermoplastic material like most influential process parameters A. K. Makawana and V. R. Patel [11] list amplitude, weld pressure and welding time. S. K. Bhudolia et al. [1] reviewed advances in ultrasonic welding of thermoplastic composites and listed welding time, frequency, amplitude and compression force as process parameters. G. Palardy and I. Fernandez Villegas [12] study the effect of flat energy directors thickness on heat generation during ultrasonic welding of thermoplastic composites and specify fixed sonotrode diameter, compression force, vibration amplitude and welding time as process parameters. T. Chinnadurai et al. [13] deal with the prediction of process parameters of ultrasonically welded joints and as parameters they investigate frequency, power, amplitude, compression force and welding time as well as three mechanical parameters (tensile strength, compressive strength and flexural strength), one physical (density) and four thermal parameters (glass transition temperature, thermal conductivity, heat deflection temperature and maximal operating temperature). S. C. Petriceanu et al. [14] use the parameters welding time, compression force, frequency and temperature when investigating the ultrasonic characterization of PVC welded materials. The best overview of the process and acoustic parameters as well as the material parameters (especially from the aspect of thermal properties) for the plunge method is given by V. N. Kmelev et al. [15], as shown in the available literature. Most of the literature refers to research into the plunge method, while the rotary sonotrode and continuous welding are poorly represented. Therefore, this paper considers the influence and interrelation of 44 parameters related to the technique of ultrasonic welding using an ultrasonic rotary sonotrode and a mathematical model of connecting parameters was established. The accuracy of the model was verified by measuring the breaking force of ultrasonically welded joints and experimentally determined welding times and power of the ultrasonic generator.2. Parameters of ultrasonic welding.

The ultrasonic welding parameters are broken down into four categories: technical parameters pertaining to the characteristics of the ultrasonic welding machine; general physical parameters pertaining to the physics of particle vibration; acoustics; and parameters pertaining to the technological process of ultrasonic welding.

The ultrasonic welding parameters used are:

(a) Parameters of the polymer material to be ultrasonically welded:

- material thickness (d, mm)
- specific material density of the polymer material to be welded (ρ_1, kgm^{-3})
- speed of ultrasound propagation in the polymer material to be welded (c_1, ms^{-1})
- acoustic damping factor of the material (μ_A, m^{-1})
- acoustic impedance of the polymer material to be welded (Z_1, kg/m^2s)

- heating the material from room temperature to melting temperature (Q_H, J)
- melting heat of the material (Q_L, J)
- (specific melting heat) latent heat of melting the material (L, J/kg)
- specific heat of the material to be welded (c, J/m^{3}K)
- initial temperature of the material before welding ((T_1, K)
- melting temperature of the material (T_2, K)
- breaking force of ultrasonic welded joints (F_p, N)

(b) Acoustic parameters:

- intensity of ultrasonic oscillations (I, W/m^2)
- coefficient of reflection of ultrasound pressures (r)
- ultrasound intensity absorption coefficient (α)
- ultrasound intensity reflection coefficient between the material to be welded and the counter roller (R)
- ultrasound intensity reflection coefficient between the ultrasonic rotary sonotrode and the material to be welded (R_1)
- ultrasound intensity absorption coefficient on the discontinuity of the sonotrode/material (α_1)
- ultrasound intensity absorption coefficient on the material/counter roller discontinuity (α_2)
- functional dependence of the decrease in ultrasound pressure in the material to be welded depending on the distance (thickness) of the material (p(x))
- ultrasound pressure (p, Pa)
- functional dependence of the decrease in ultrasound intensity in the material with distance (thickness) of the material (I(x), W/m^2)
- coefficient of reflection of ultrasound intensities on the discontinuity (R)

(c) Technological parameters (depending on the machine):

- frequency of ultrasonic vibrations of the sonotrode (f, Hz)
- declared electrical power of the machine's ultrasonic generator (P_d, W)
- electrical power of the ultrasonic generator of the machine at which the maximum breaking force of the ultrasonic welded joint appears (P_{dmax}, W)
- effective power of the ultrasonic rotary sonotrode (P_α, W)
- force with which rotary ultrasonic sonotrodes act on two layers of thermoplastic polymer materials with total thickness (P_s, N)
- specific density of sonotrode material (ρ_0, kg/m^3)
- speed of ultrasound propagation in the sonotrode (c_0, ms^{-1})
- sonotrode vibration amplitude (A_0, m)
- acoustic impedance of the ultrasonic rotary sonotrode (Z_0, kg/m^2 s)
- specific material density from which the counter-roller is made (ρ_2, kgm^{-3})
- speed of ultrasound propagation of the backing material for welding (c_2, ms^{-1})
- acoustic impedance of the backing material (table) for welding (Z_3, kg/m^{2}s)
- sonotrode ultrasound intensity (I_0, W/m^2)
- radius of the ultrasonic rotary sonotrode (r_s, m)
- width of the ultrasonic rotary sonotrode (w_s, m)
- length of the imprint of the rotary sonotrode on the material (l_s, m)
- angular velocity of the rotation of the ultrasonic sonotrode (ω_s, rad/s). If the linear velocity of the sonotrode edge and the radius of the rotary ultrasonic sonotrode are given, the angular velocity is calculated by the expression $\omega_s = v_s/r_s$
- linear speed of the edge of the ultrasonic rotary sonotrode, i.e., speed of the movement of the workpiece v_s, cm/s). If the angular velocity and the radius of the rotary ultrasonic sonotrode are given, the linear velocity is calculated by the expression $v_s = \omega_s.r_s$.

- ultrasonic welding time depending on the ratio of the length of the imprint of the ultrasonic rotary sonotrode and the linear speed of the sonotrode edge, i.e., the speed of the workpiece movement (t_s, s)
- ultrasonic welding time calculated according to the mathematical model (t_m, s)
- delay time of the start of welding due to heating the beginning of the welded joint (t_{dy}, s).

2. Acoustic Mathematical Model of Ultrasonic Welding Time

Figure 1 shows a rotary ultrasonic sonotrode acting with force F when welding two layers of thermoplastic polymers with a total thickness of 2d. The ultrasonic rotary sonotrode has a specific density of the material ρ_0 from which it is made and the speed of sound waves c_0 in it. The polymer material has a specific material density ρ_1 and the speed of sound waves c_1 in it. At the bottom of the welder there is a counter-roller that has a specific density of the material ρ_2 from which it is made and the speed of sound waves c_2 within it. Ultrasonic vibrations of intensity I_0 occur in the rotary ultrasonic sonotrode, which are intended to induce heat into the polymer material. Since reflection occurs on each discontinuity of acoustic impedances encountered by an ultrasonic wave, an ultrasonic wave of intensity I_0 is also partially reflected on the discontinuity (sonotrode/material) with an intensity I_{r0}, while a larger part penetrates the material with an intensity of I_1. Due to the absorption of the material and the conversion of the ultrasonic wave into heat, the intensity of the wave weakens, so that it takes on the intensity of I_{12} at the contact surface between the lower part of the material and the counter roller. Since it is the second discontinuity (material/counter roller), there is a reflection of intensity I_{r1}, which returns part of the ultrasonic energy into the material, while the other part with an intensity of I_2 goes into the counter roller, representing a loss of ultrasonic energy.

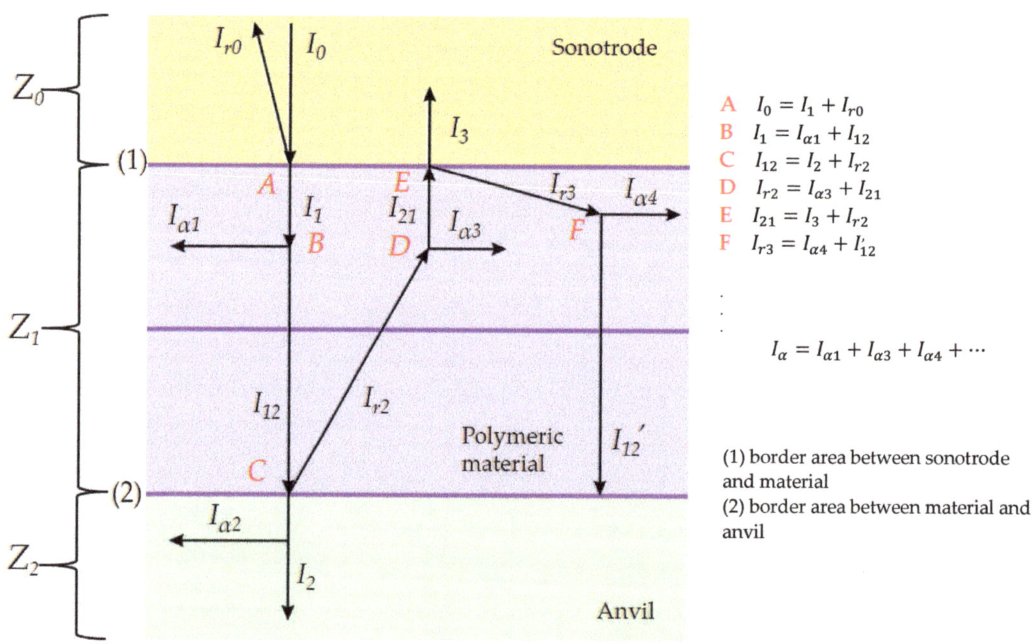

Figure 1. Machine components and acoustic parameters used in the ultrasonic welding process utilizing the rotary sonotrode ultrasonic welding machine.

Due to material absorption, the counter roller's reflected ultrasonic wave, initially of intensity I_{r1}, gradually becomes of intensity I_{21}. Back-reflection of I_{r2} on the discontinuity (material/sonotrode) causes a portion of the ultrasonic energy to return to the material and a portion of the ultrasonic wave with an intensity of I_3 to reach the sonotrode.

The intensity of the ultrasonic oscillations of the sonotrode I_0, i.e., the average power of the ultrasonic wave, as determined by E. Dieulasaint and D. Royer [16], can be expressed by Equation (1):

$$I_0 = 2 \cdot \pi^2 \cdot f^2 \cdot A_0^2 \cdot Z_0 \quad (1)$$

The acoustic impedance Z_0 is defined as the product of the specific density of the material ρ_0 from which the ultrasonic rotary sonotrode is made and the speed of the sound wave c_0 in it in accordance with Equation (2):

$$Z_0 = \rho_0 \cdot c_0 \quad (2)$$

Similarly, the acoustic impedances of the polymer material to be welded Z_1 and the counter roller Z_2 are defined according to equations:

$$Z_1 = \rho_1 \cdot c_1 \quad (3)$$

$$Z_2 = \rho_2 \cdot c_2 \quad (4)$$

The coefficient of reflection of ultrasound pressures (ratio of RMS values of reflected p_r and direct ultrasound pressure p_d) is defined by Equation (5):

$$r = \frac{p_r}{p_d} \quad (5)$$

The ultrasound intensity absorption coefficient (ratio of absorbed $I\alpha$ and direct I_d of ultrasound intensity) is defined by Equation (6):

$$\alpha = \frac{I_\alpha}{I_d} \quad (6)$$

Equation (7) describes the relationship between the coefficient of absorption of ultrasound intensities and the coefficient of reflection of ultrasound pressures (7):

$$\alpha = 1 - r^2 \quad (7)$$

The sound intensity reflection coefficient (ratio of reflected I_r and direct I_d of ultrasound intensity) is defined by Equation (8):

$$R = \frac{I_r}{I_d} = r^2 \quad (8)$$

The functional dependence of a decrease in ultrasound pressure in a material depending on material distance is defined by Equation (9):

$$p(d) = p \cdot e^{-\mu_A \cdot d} \quad (9)$$

According to the relation (10), the ultrasound intensity is inversely proportional to the ultrasonic pressure.

$$I \sim p^2 \quad (10)$$

The functional dependence of a decrease in ultrasound intensity in a material with distance (thickness) is defined by Equation (11):

$$I(d) = I \cdot e^{-2 \cdot \mu_A \cdot d} \quad (11)$$

The coefficient of reflection of ultrasound intensities on the discontinuity sonotrode/material, designated as (1) in Figure 2, taking into account the acoustic impedances of sonotrode and material Z_0 i and material Z_1 is defined by Equation (12):

$$R_1 = \left(\frac{Z_0 - Z_1}{Z_0 + Z_1}\right)^2 \tag{12}$$

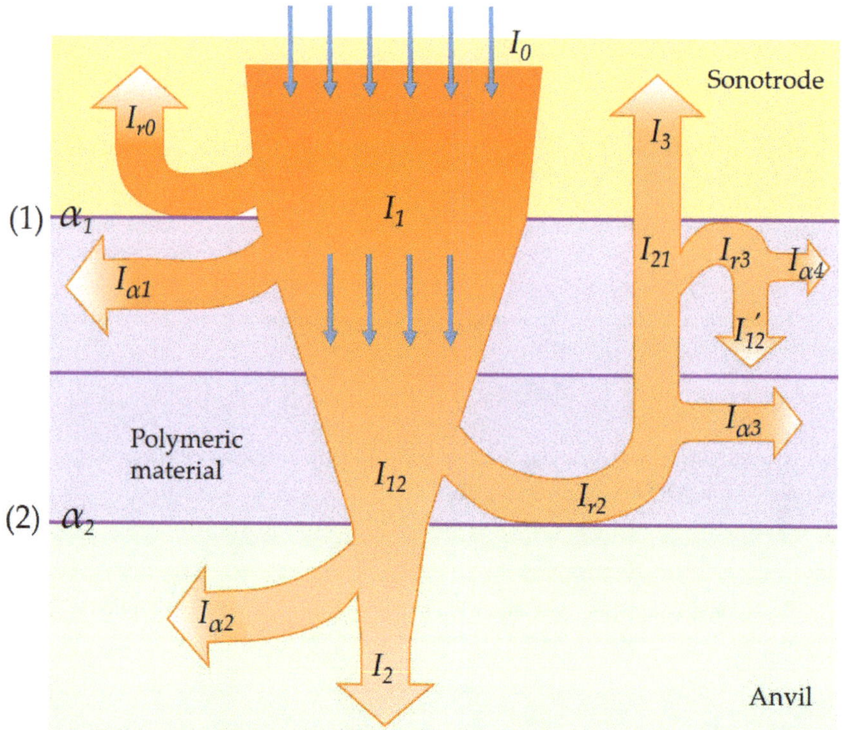

Figure 2. A symbolic author's presentation of the intensity, propagation, reflection and absorption of ultrasonic waves, suitable for production engineering.

The coefficients of reflection of ultrasound intensities on the discontinuity material/counter roller, designated as (2) in Figure 1, taking into account the acoustic impedances of the material Z_1 and the counter roller Z_2 is defined by Equation (13):

$$R_2 = \left(\frac{Z_1 - Z_2}{Z_1 + Z_2}\right)^2 \tag{13}$$

On discontinuities the ultrasound intensity absorption coefficient on discontinuity designated (1) of acoustic impedances of sonotrode and material 1 (name of material) can be expressed by Equation (14):

$$\alpha_1 = 1 - R_1 \tag{14}$$

and on the discontinuity designated as (2) in Figure 2 by Equation (15):

$$\alpha_2 = 1 - R_2 \tag{15}$$

The ultrasound intensity that penetrates the polymer material can be defined by equation:

$$I_1 = \alpha_1 \cdot I_0 = (1 - R_1) \cdot I_0 \tag{16}$$

Due to the absorption of the energy of the ultrasonic waves $I_{\alpha 1}$ in the thermoplastic polymer material, the temperature of the material rises and the ultrasonic wave penetrating through the material weakens in intensity, so that it drops from the initial I_1 to the intensity I_{12} at the bottom of the layer of the welded material. In this way, the absorption value of the injected ultrasonic wave from the sonotrode can be calculated in the polymer material:

$$I_{\alpha 1} = I_1 - I_{12} \tag{17}$$

The absorption of the energy of the ultrasonic waves $I_{\alpha 1}$, which is converted into heat, is shown in Figure 1.

The intensity of the ultrasound wave on the bottom side of the layer of the material to be welded takes the value:

$$I_{12} = I_1 \cdot e^{-2 \cdot \mu_A \cdot d} \tag{18}$$

by inserting the expressions (17) and (18), the following results:

$$I_{\alpha 1} = I_1 - I_1 \cdot e^{-2 \cdot \mu_A \cdot d} \tag{19}$$

$$I_{\alpha 1} = I_1 \left(1 - e^{-2 \cdot \mu_A \cdot d}\right) \tag{20}$$

The intensity of the reflected wave I_{r1} on the discontinuity (2) (material/counter roller) returning to the polymer material is defined by the equation:

$$I_{r1} = R_2 \cdot I_{12} \tag{21}$$

The aforementioned reflected wave attenuates on its path between the discontinuity (2) and the discontinuity (1) and has the amount I_{21} at the end of its path, near the top of the material layer:

$$I_{21} = I_{r1} \cdot e^{-2 \cdot \mu_A \cdot d} \tag{22}$$

The following set of equations can be stated because the weakening of the reflected ultrasonic wave is transformed into heat that further warms the polymer material:

$$I_{\alpha 3} = I_{r1} - I_{21}$$

$$I_{\alpha 3} = I_{r1} - I_{r1} \cdot e^{-2 \cdot \mu_A \cdot d}$$

$$I_{\alpha 3} = R_2 \cdot I_{12} - R_2 \cdot I_{12} \cdot e^{-2 \cdot \mu_A \cdot d}$$

$$I_{\alpha 3} = R_2 \cdot I_{12} \cdot \left(1 - e^{-2 \cdot \mu_A \cdot d}\right)$$

$$I_{\alpha 3} = R_2 \cdot I_1 \cdot e^{-2 \cdot \mu_A \cdot d} \cdot \left(1 - e^{-2 \cdot \mu_A \cdot d}\right) \tag{23}$$

It can be deduced that the sum of the heat created by the advancing wave from the sonotrode to the counter roller and the reflected wave from the counter roller back to the sonotrode represents the total heat I_1 developed in the polymer material:

$$I_\alpha = I_{\alpha 1} + I_{\alpha 3} \tag{24}$$

Figure 2 illustrates the author's symbolic representation of the ultrasonic wave's intensity, propagation, reflection, and absorption in order to highlight the many effects on the thermoplastic polymer materials it is used to weld. The aforementioned presentation aims to make it easier to comprehend the mathematical equations linking the dependencies of the technical and technological parameters of polymer ultrasonic welding.

Figure 2 shows the largest ultrasound wave generated in the sonotrode with intensity I_0, which is directed towards the thermoplastic polymer material according to its effect. When the ultrasound wave reaches the discontinuity (1) at the boundary of the sonotrode and the upper layer of the material being welded, a reflection occurs, whereby a part of the ultrasonic wave I_{r0} is reflected back into the sonotrode. The weakened ultrasonic wave with intensity I_1 continues to penetrate the polymer material. As the ultrasonic wave penetrates the polymer material, the ultrasonic energy $I_{\alpha 1}$ is absorbed, causing the intensity of the ultrasonic wave to decrease so that it assumes intensity I_{12} immediately adjacent to the second discontinuity (2) at the boundary between the bottom layer of the material and the counter roller. When the ultrasonic wave reaches the discontinuity (2), a back-reflection occurs, where part of the ultrasonic wave of intensity I_{r1} is reflected back into the polymer material and the other part of intensity $I_{\alpha 2}$ is transferred to the counter roller, where part of the energy of the ultrasonic wave$_2$ is irreversibly lost. The ultrasonic wave reflected from the discontinuity (2) propagates towards the sonotrode, where it weakens because it returns a large part of its energy to the material $I_{\alpha 3}$. The weakened reflected wave finally reaches the sonotrode, i.e., the discontinuity (1), so that part of the wave is reflected back into the material with an intensity of I_{r2}, while a part is transmitted to the sonotrode I_{r3}.

The influence of the penetration of the ultrasonic waves into the counter roller I_2 and into the sonotrode I_3 as well as the influence of the back reflection of the reflected wave Ir$_3$ are neglected in this analysis, as their influence is small when the influence of the other parameters is taken into account.

The following set of equations, which apply to a layer of material in the welding process, are generated by entering Equations (20) and (23) into Equation (24), followed by introducing expression (1) into expression (24):

$$I_\alpha = I_1 \cdot \left(1 - e^{-2 \cdot \mu_A \cdot d}\right) + R_2 \cdot I_1 \cdot e^{-2 \cdot \mu_A \cdot d} \cdot \left(1 - e^{-2 \cdot \mu_A \cdot d}\right)$$

$$I_\alpha = I_1 \cdot \left(1 - e^{-2 \cdot \mu_A \cdot d} + R_2 \cdot e^{-2 \cdot \mu_A \cdot d} - R_2 \cdot e^{-4 \cdot \mu_A \cdot d}\right)$$

$$I_\alpha = (I_1 - R_1) \cdot I_0 \cdot \left(1 - e^{-2 \cdot \mu_A \cdot d} + R_2 \cdot e^{-2 \cdot \mu_A \cdot d} - R_2 \cdot e^{-4 \cdot \mu_A \cdot d}\right)$$

$$I_\alpha = I_0 \cdot (I_1 - R_1) \cdot \left(1 - e^{-2 \cdot \mu_A \cdot d} + R_2 \cdot e^{-2 \cdot \mu_A \cdot d} - R_2 \cdot e^{-4 \cdot \mu_A \cdot d}\right)$$

$$I_\alpha = 2 \cdot \pi^2 \cdot f^2 \cdot A_0^2 \cdot Z_0 \cdot (I_1 - R_1) \cdot \left(1 - e^{-2 \cdot \mu_A \cdot d} + R_2 \cdot e^{-2 \cdot \mu_A \cdot d} - R_2 \cdot e^{-4 \cdot \mu_A \cdot d}\right) \tag{25}$$

However, as the overall thickness of the material is doubled during ultrasonic welding of the two material layers, the following is applicable:

$$I_\alpha = 2 \cdot \pi^2 \cdot f^2 \cdot A_0^2 \cdot Z_0 \cdot (I_1 - R_1) \cdot \left(1 - e^{-2 \cdot \mu_A \cdot 2 \cdot d} + R_2 \cdot e^{-2 \cdot \mu_A \cdot 2 \cdot d} - R_2 \cdot e^{-4 \cdot \mu_A \cdot 2 \cdot d}\right)$$

$$I_\alpha = 2 \cdot \pi^2 \cdot f^2 \cdot A_0^2 \cdot Z_0 \cdot (I_1 - R_1) \cdot \left(1 - e^{-4 \cdot \mu_A \cdot d} + R \cdot e^{-4 \cdot \mu_A \cdot d} - R \cdot e^{-8 \cdot \mu_A \cdot d}\right) \tag{26}$$

The ultrasonic power of P_α developed in the polymer material is calculated by the product of the intensity of ultrasonic waves that are converted into the heat of the polymer material and the area of action of the sonotrode consisting of the width of the sonotrode w_s and the length of the sonotrode imprint on the polymer material l_s:

$$P_\alpha = I_\alpha \cdot w_s \cdot l_s \tag{27}$$

The following equation is created if Equation (26) is inserted into Equation (27):

$$P_\alpha = 2 \cdot \pi^2 \cdot f^2 A_0^2 \cdot Z_0 \cdot w_s \cdot l_s \cdot (1 - R_1) \cdot \left(1 - e^{-4 \cdot \mu_A \cdot d} + R \cdot e^{-4 \cdot \mu_A \cdot d} - R \cdot e^{-8 \cdot \mu_A \cdot d}\right) \tag{28}$$

When ultrasonic energy is first injected into a polymer material, the temperature of the substance starts to rise as the energy is used to raise the material's temperature from its initial temperature to the melting point. The density of the material ρ , the thickness of the two material layers with individual material thickness d, the width of the sonotrode w_s, the length of the sonotrode imprint on the polymer material l_s, the specific heat of the material c, and the difference between the melting temperature and the initial temperature are the factors that determine the energy needed to reach the melting temperature of the ultrasonic material:

$$Q_H = 2 \cdot \rho \cdot d \cdot w_s \cdot l_s \cdot c \cdot (T_2 - T_1) \tag{29}$$

The latent melting heat of the welded polymer material Q_L is equal to the product of the density of the material ρ, the thickness of the two layers of material with individual thickness d, the width of the sonotrode w_s, the length of the imprint of the sonotrode on the polymer material l_s and the latent heat of the material L:

$$Q_L = 2 \cdot \rho \cdot d \cdot w_s \cdot l_s \cdot L \tag{30}$$

The total heat Q_T required for welding polymer materials is the sum of specific heating and latent melting heat, so the addition of Equations (29) and (30) gives Equation (31):

$$Q_T = 2 \cdot \rho \cdot w_s \cdot l_s \cdot [c \cdot (T_2 - T_1) + L] \tag{31}$$

If one takes into account the well-known physical equation that the energy $Q_T = P_\alpha \cdot t$, then $t = Q_T / P_\alpha$. Thus, if Equations (31) and (28) are inserted into this expression, an equation is obtained for calculating the required welding time t during the action of ultrasound on thermoplastic polymer materials:

$$t = \frac{\rho \cdot [c \cdot (T_2 - T_1) + L]}{\pi^2 \cdot f^2 A_0^2 \cdot Z_0 \cdot (1 - R_1) \cdot \left(1 - e^{-4 \cdot \mu_A \cdot d} + R \cdot e^{-4 \cdot \mu_A \cdot d} - R \cdot e^{-8 \cdot \mu_A \cdot d}\right)} \tag{32}$$

By omitting some action parameters (lateral thermal expansion into a polymer material that is insignificant given the poor thermal conductivity of polymer materials, neglecting the thermal effect at the second reflection from the discontinuity of intensity $I_{\alpha 4}$, etc.), Equation (32) is still of high accuracy in terms of practical application. Equation (32) has a complicated structure that indicates there are many factors that affect how polymer materials weld. The most important parameters are the welding times and the power of the ultrasonic generator, expressed by the amplitudes of the ultrasonic rotary sonotrode. These are also the most frequently set parameters in ultrasonic welding. The aforementioned equation and all of its derivations, which completely describe and reveal its origin, very thoroughly reveal the intricate interdependencies of the technical and technological parameters of the ultrasonic welding of polymer materials.

Based on experience and experiments, it was found that the most important parameters of the comprehensive model are the declared power of the ultrasonic generator and the welding times, while other parameters are required to achieve a certain accuracy of the acoustic mathematical model of ultrasonic welding time.

3. Measuring Methods and Materials

From mathematical expression (32) it can be seen that for a certain material, which is also welded for a certain machine, the amplitude of the ultrasonic rotary sonotrode A_0 and the welding time are parameters that can change. Other parameters do not change during joining in this case. The machine manufacturers do not specify the parameter value for the amplitude, but allow that the power of the ultrasonic generator P_d can be adjusted. Therefore, it is important to check the value of the amplitude of the ultrasonic rotary sonotrode at the declared power of the ultrasonic generator.

The vibration amplitudes of the ultrasonic rotary sonotrode A_0 and their dependence on the indicated declared power of the ultrasonic generator, which is typically displayed on the microcomputer screen of the ultrasonic machine, were measured to verify the mathematical model of ultrasonic welding time presented in Section 3 of this paper and in mathematical expressions (1) to (32). The declared power of the ultrasonic generator have been determined for Pfaff Model 8310 [17] ultrasonic machine with rotary sonotrode for welding polymer materials, Figure 3. A rotary sonotrode, Figure 4, with a welding speed of 0.6 to 13.6 m/min is a component of the ultrasonic machine. The ultrasonic generator has an adjustable power range of 200 to 400 W and runs at a frequency of 35 kHz. It sends the vibrations to a circular sonotrode with a 105 mm diameter and a 10 mm width.

Test of declared power of the ultrasonic generator and appropriate vibration amplitudes for this power were made using an Agilent Technologies InfiniVisions MSO-X 3024A oscilloscope [18] and an MTI-2100 photonic sensor device [19].

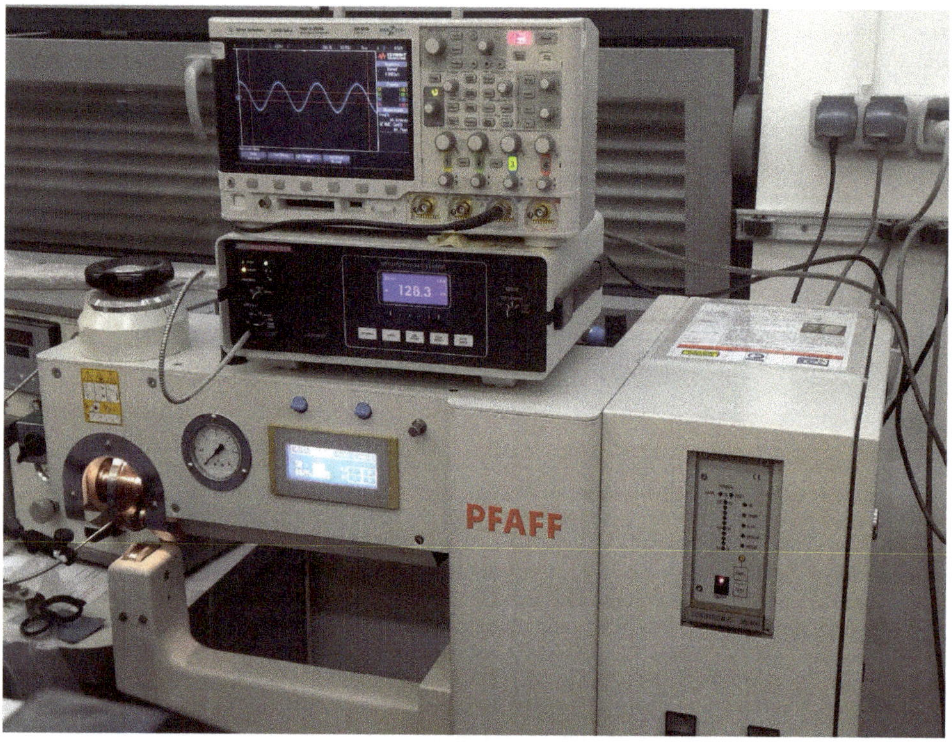

Figure 3. A measuring setup on the machine for welding polymer materials.

The photon sensor MTI-2100, a dual-channel optical measurement system that carries out non-contact measurements of displacements and vibrations, was utilized in this work. Because the MTI-2100 employs fiber optic technology, which implies that it does not subject the object being measured to stress and is not impacted by magnetic and electric forces, it was selected. At frequencies ranging from DC to more than 150 kHz, displacements in the 0.25 nm to 5.08 mm range can be measured. The maximum linear range in normal winning mode is in the order of 1 million (25 μm). The frequency response ranges from DC to 100 kHz. The typical normal sensitivity of probes is to the order of 0.025 μm/mV. Optically coupled neighboring pairs that send and receive fiber light are used by optical fiber displacement sensors. The interplay between the field of light transmission, or the original fibers, and the field of view of the receiving or detector fibers serves as the foundation for the research. At contact or zero spacing, the majority of the light from the original fibers is reflected back into the same fiber, sending just a little amount of light to the receiving fibers and resulting in a zero-power signal. With an output signal that has the following characteristics: −10 V DC to 10 V DC, 51 output impedance, stability at 12 h 1 °C, displacement of less than 1.0% of full scope, and stability at a temperature between 16 °C and 35° [18], displacement tests are achievable.

Figure 4. A rotary sonotrode and counter roller for the ultrasonic welding of polymer materials.

The light source and light sensor are the two components of the displacement measurement system that is based on the measurement of light intensity (detector). The relocation results in a change in the optical power that reaches the detector in the most basic systems. Infrared LEDs and LASERs with greater ranges are typically used as visible light source. Such arrangements are advised for modest displacements because to their good linearity and steep upstream wing curve. To prevent cross-consolidation between the transmitting and receiving fibers, the fibers must be of the total internal reflection variety. The sensitivity depends on where both fiber types are placed in the beam. Even though they are affected by the shape of the beam-interrupting item, the cross-section of the detector, and the spatial distribution of the intensity of the light emitted by the source, the relationship between the light reaching the detector and the displacement is linear [20]. In this way, the diagram dependence of the vibration amplitude of the rotary ultrasonic sonotrode on the declared power of the ultrasonic generator is defined. The amplitude values determined in this way are used to calculate the welding times according to the mathematical model.

In this paper, the quality of the welded joint was investigated in this paper by creating 42 groups of specimens welded at 7 welding speeds (i.e., joint times that depend on the circumferential speed of the sonotrode) and 6 declared electrical power of the ultrasonic generator. In this way, it is possible to determine the functional dependence of breaking forces of the ultrasonically welded joint on the declared power for each circumferential speed, i.e., welding time. From the diagram, the optimum welding parameters (generator power and circumferential speed of the ultrasonic rotary sonotrode) can be determined when the cutting force is maximum. The optimum parameter values can be useful for checking the values of the joint parameters according to the model and experiment.

A Sauter tensile tester, model HF 500 [21], with a breaking strength of up to 500 N and a resolution of 0.1 N, was used to explore the impact of ultrasonic welding parameters on the breaking force of ultrasonic welds, which is dependent on the quantity of ultrasonic energy supplied. Figure 5a shows the Sauter Electrical Horizontal Test Stand THM 500N500N with the tearing strength tester mounted on it [22]. The test stand includes an adjustable elongation speed and automatic and manual control options. The strength tester's test clamps, which are used to determine the test specimen's width, have a 15 mm width. The test specimen was sandwiched between the strength tester's movable clamp and the Test Stand's fixed clamp, as shown in Figure 5b.

(**a**) (**b**)

Figure 5. The measuring system for measuring breaking forces of an ultrasonic welded joint (**a**), test clamps with a test specimen after tearing (**b**).

The test material is a PVC foil with a thickness of 20 μm. The specific heat of the tested material c is 1×10^3 J/mm^{3}K, at the melting temperature T2 of 485 K and the ambient temperature T_1 of 289 K. The specific melting heat L is 163×10^3 J/kg, and the specific density of the polymer material ρ is 1.4×10^3 kg/m^3.

PVC foil was used to create 15×50 mm^2 measuring specimens. The aforementioned ultrasonic machine, which has an adjustable electrical power of the ultrasonic generator of 50 to 100%, or converted to a power of 200 to 400 W, was used to weld the test specimens. For this experiment, six power levels of 200 W, 244 W, 280 W, 320 W, 356 W, and 400 W were chosen. Seven welding speeds were also used, including 0.077 m/s, 0.097 m/s, 0.113 m/s, 0.125 m/s, 0.147 m/s, 0.167 m/s, and 0.227 m/s. To calculate the breaking force of the ultrasonic welds for each of the specified materials, 42 groups of 20 test specimens were created in line with the aforementioned information and evaluated on a strength tester.

The acoustic impedance of the sonotrode Z_0 was 4.1×10^7 kg/m^2 s, the acoustic impedance of the material to be welded (acoustic impedance of the material), Z_1, was 3.2×10^6 kg/m^{2}s. The acoustic damping factor of the material μ_A was 0.37 m^{-1}.

The ultrasound intensity reflection coefficient between the material to be welded and the counter roller R and the ultrasound intensity reflection coefficient between the sonotrode and the material to be welded R_1 amount to 0.73.

4. Results

This chapter includes results from measurements of the rotary ultrasonic sonotrode's vibration amplitude as a function of the ultrasonic generator's declared electrical power, the breaking force of ultrasonic welded joints for 42 groups of test specimens at various strengths and welding speeds, the experimental data used to validate the mathematical model, and an assessment of the model's effectiveness.

4.1. Results of Analysing the Rotary Sonotrode's Vibration Amplitude in Relation to the Ultrasonic Generator's Set Electrical Power

The creation of a technique to detect the vibration amplitudes of the rotary sonotrode using a photonic sensor device is one of the contributions of this work, which is covered in Section 4 of this paper. Six specific powers of the ultrasonic generator were used to measure the oscillation amplitude of the rotary sonotrode (200 W, 244 W, 280 W, 320 W, 356 W and 400 W). The outcomes are displayed in Figure 6.

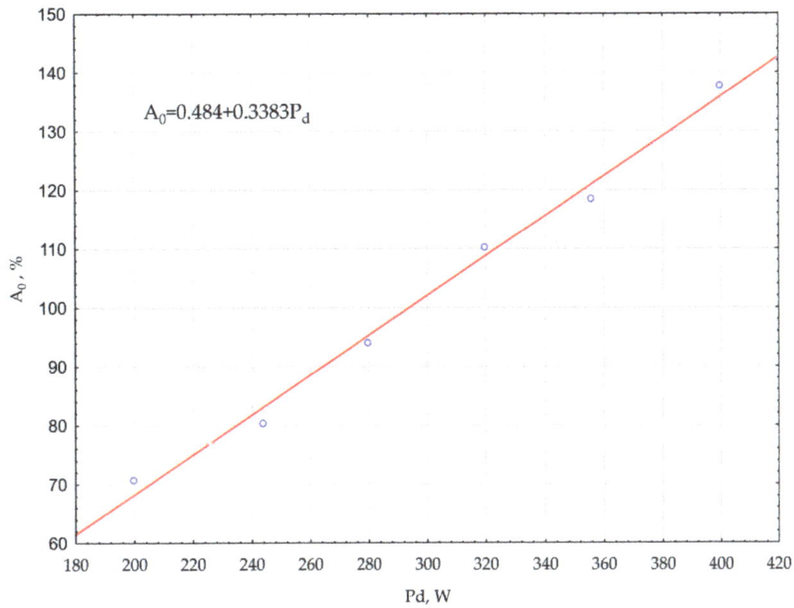

Figure 6. Dependence of the vibration amplitude of the rotary ultrasonic sonotrode on the declared power of the ultrasonic generator.

The best connection between functional dependencies is in the form of a straight line with the equation $A_0 = 0.484 + 0.3383 \cdot P_d$.

4.2. Results of Testing Specimens of Ultrasonic Welds for Breaking Forces

The breaking force measurements were carried out for 42 groups of specimens welded at different welding speeds (0.077 m/s, 0.097 m/s, 0.113 m/s, 0.125 m/s, 0.147 m/s, 0.167 m/s, 0.227 m/s) and different declared powers of the ultrasonic generator (200 W, 244 W, 280 W, 320 W, 356 W and 400 W), Table 1.

Section 4 describes additional measuring circumstances. Table 1 displays the results of the breaking force measurements and standard deviations of the ultrasonic weld specimens for 42 groups of test specimens. Grades for visual observation of ultrasonic welds prior to tearing are provided in Table 1.

A. Weld with a nice appearance, Figure 7a
B. Weld with scarcely noticeable damage symptoms, Figure 7b
C. Weld with substantial weld damage, Figure 7c.

Three visual observation grades of the ultrasonic weld following tearing are also shown in Table 1:

D. Separation of the material layers at the weld
E. Material tearing inside the weld
F. Material tearing along the weld's edge.

Table 1. The results of breaking force values and standard deviations of specimens of ultrasonic welds and visual observation grades of ultrasonic welds before and after tearing.

v_s, m/s	Testing Elements	P_d, W					
		200	244	280	320	356	400
0.077	Visual observation grades of ultrasonic welds before tearing	A	A	A	C	C	C
	Breaking force of ultrasonic welded joint F [N]	33.80	30.80	23.90	-	-	-
	Standard deviation	4.66	6.2	3.12	-	-	-
	Visual observation grades of ultrasonic welds after tearing	D, E	E	E	-	-	-
0.097	Visual observation grades of ultrasonic welds before tearing	A	A	A	B	C	C
	Breaking force of ultrasonic welded joint F [N]	28.24	32.24	31.21	27.75	-	-
	Standard deviation	1.46	4.02	4.09	3.01	-	-
	Visual observation grades of ultrasonic welds after tearing	D	D, E	E	E, F	-	-
0.113	Visual observation grades of ultrasonic welds before tearing	A	A	A	B	C	C
	Breaking force of ultrasonic welded joint F [N]	23.14	31.00	32.34	26.00	-	-
	Standard deviation	2.95	3.00	3.95	2.10	-	-
	Visual observation grades of ultrasonic welds after tearing	D	D	D, E	E, F	-	-
0.125	Visual observation grades of ultrasonic welds before tearing	A	A	A	A	C	C
	Breaking force of ultrasonic welded joint F [N]	21.20	30.15	33.50	24.14	-	-
	Standard deviation	5.49	2.41	2.63	3.07	-	-
	Visual observation grades of ultrasonic welds after tearing	D	D	D, E	F	-	-
0.147	Visual observation grades of ultrasonic welds before tearing	A	A	A	A	C	C
	Breaking force of ultrasonic welded joint F [N]	15.32	24.66	28.48	30.94	-	-
	Standard deviation	2.87	2.05	3.12	2.64	-	-
	Visual observation grades of ultrasonic welds after tearing	D	D	D	D	-	-
0.167	Visual observation grades of ultrasonic welds before tearing	A	A	A	A	C	C
	Breaking force of ultrasonic welded joint F [N]	12.30	22.12	26.00	28.60	-	-
	Standard deviation	1.38	1.62	2.93	4.01	-	-
	Visual observation grades of ultrasonic welds after tearing	D	D	D	D	-	-
0.227	Visual observation grades of ultrasonic welds before tearing	-	-	A	A	A	A
	Breaking force of ultrasonic welded joint F [N]	-	-	5.40	14.33	21.40	26.30
	Standard deviation	-	-	3.41	3.21	2.91	3.12
	Visual observation grades of ultrasonic welds after tearing	-	-	D	D	D	D

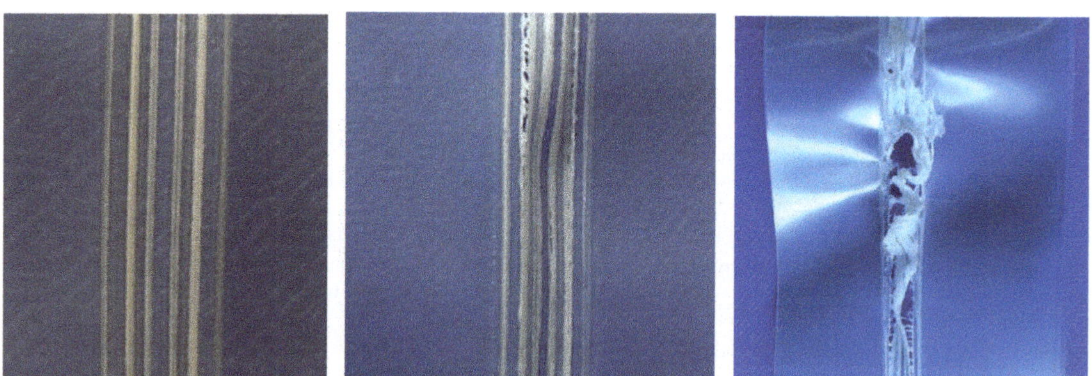

(a) (b) (c)

Figure 7. *Cont.*

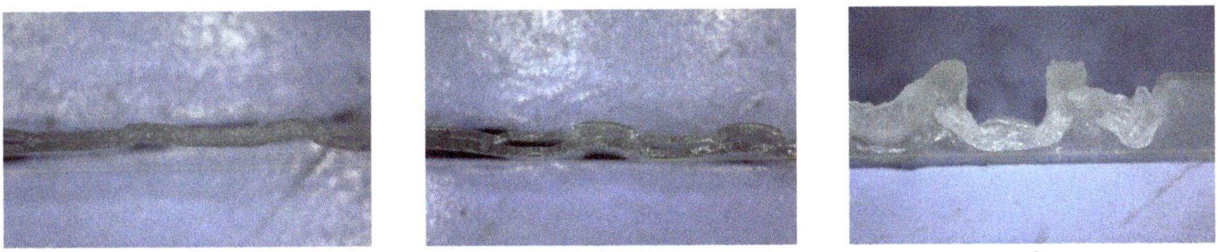

(d) (e) (f)

Figure 7. The visual appearance of a weld with a nice appearance (**a**) and its cross-sectional view (**d**), weld with scarcely noticeable damage symptoms (**b**) and its cross-sectional view (**e**), weld with substantial weld damage (**c**) and and its cross-sectional view (**f**).

Based on the findings in Table 1, Figure 8 was constructed to show the functional relationships between the breaking forces of ultrasonic welds and the declared electric power of the ultrasonic generator and the speed at which the specimens were welded using an ultrasonic rotary sonotrode.This picture is crucial for figuring out the maximum breaking forces at which the supplied ultrasonic energy causes the weld to break most violently.

Figure 8 shows that four maxima were reached:

- for a welding speed of 0.077 m/s for the declared power of the ultrasonic generator of 200 W
- for a welding speed of 0.097 m/s for the declared power of the ultrasonic generator of 250 W
- for a welding speed of 0.113 m/s for the declared power of the ultrasonic generator of 268 W
- for a welding speed of 0.125 m/s at the declared power of the ultrasonic generator of 272 W.

All maximum values reached range from 32 to 34 N breaking force.

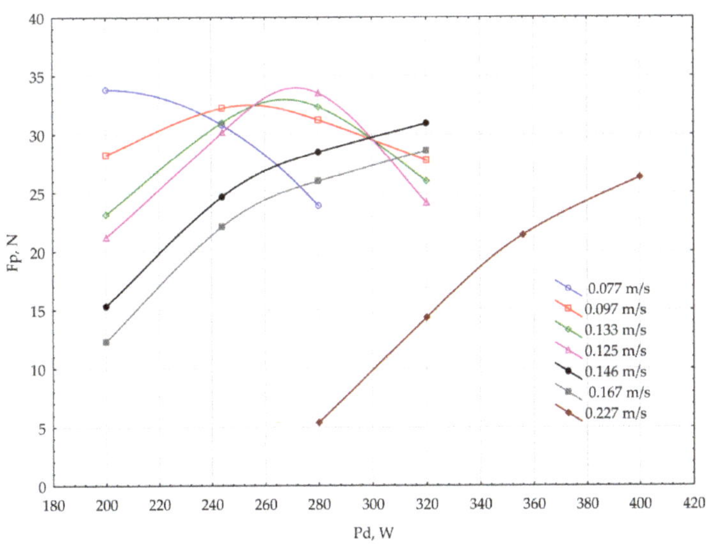

Figure 8. A diagram showing the functional relationships between the breaking forces of ultrasonic welds and the speed at which specimens are welded using a rotary ultrasonic sonotrode in relation to the declared electrical power of the ultrasonic generator.

4.3. Verification of Mathematical Model Results

The maximum breaking forces of the test specimens for four of the highest welding speeds—0.077 m/s, 0.097 m/s, 0.113 m/s, 0.125 m/s, 0.147 m/s, 0.167 m/s, and 0.227 m/s—are significant for the mathematical model's validation. These results are presented in Fig. 6. For each maximum, the vibration amplitude must be read based on the ultrasonic generator's declared power, which is 200 W, 250 W, 268 W, or 272 W.

Using the expression according to Figure 6, $A_0 = 0.484 + 0.3383 \cdot P_d$, for the declared power of the ultrasonic generator of 200 W, the oscillation amplitude of the rotary ultrasonic sonotrode was determined to be 68.1 μm, for a power of 250 W the amplitude was 85.1 μm, for a power of 268 W the amplitude was 91.1 μm and for a power of 272 W the amplitude was 92.5 μm.

In this way, the exact values of the vibrations attained by the ultrasonic rotary sonotrode were determined as a function of the declared electrical power by the measurement method used. From these data, the effective power of the rotary ultrasonic sonotrode can be calculated using expression (28) of the mathematical model presented. Thus, for the oscillation amplitude of the ultrasonic rotary sonotrode of 68.1 μm an effective power of 126.5 W was determined, for an amplitude of 85.1 μm, an effective power of 197.5 W, for an amplitude of 91.1 μm, an effective power of 226.3 W, and for an amplitude of 92.5 μm, an effective power of 233.4 W, Table 2.

Table 2. Values of the vibrations, effective power, speed and welding time achieved by the ultrasonic rotary sonotrode depend on the specified electrical power.

P_{dmax}, W	A_0, μm	P_{α}, W	v_s, m/s	t_m, s	t_s, s	Δ ($t_m - t_s$), s
200	68.1	126.5	0.077	0.288	0.259	0.029
250	85.1	197.5	0.097	0.184	0.206	−0.022
268	91.1	226.3	0.113	0.161	0.177	−0.016
272	92.5	233.4	0.125	0.156	0.160	−0.004

The workpiece movement speed vs. and the calculated welding time s at various speeds are shown in Table 2 in accordance with expression (32) of the mathematical model t_m that was introduced in the paper's introduction. Additionally, the actual times t_s are displayed, which are derived from the linear speed of the rotary sonotrode circumference ($t_s = l_s/v_s$) and the 0.02 m imprint value of the rotary sonotrode.

5. Discussion

The article lists 44 crucial process variables for welding polymer materials. There are three groups of parameters: parameters specific to the polymer material being ultrasonically welded (12 parameters), general acoustic and electroacoustic parameters (11 parameters), and technical parameters specific to the ultrasonic machine (21 parameters). Important acoustic parameters are also shown, including reflections and absorptions between the rotating counter roller, the polymer material, and the rotary ultrasonic sonotrode. This presentation served as the basis for the methodical creation of a mathematical model that may be utilized to determine several ultrasonic welding parameters. Additionally, in developing the model, consideration is given to how ultrasonic wave strength, propagation, reflections, and absorption relate to production engineering.

The approach described in this study uses a photon sensor to measure the vibration amplitudes of a rotary ultrasonic sonotrode while it is operating. The aforementioned measurement technique proved to be extremely accurate and well accepted during the experimental portion of this work. Using this technique, the ultrasonic rotary sonotrode vibration amplitudes can be modified on the ultrasound machine's microcomputer screen for six different values of the ultrasonic generator's stated power (200 W, 244 W, 280 W, 320 W, 356 W, and 400 W). As a result, a calibrated diagram was created, as seen in Figure 6.

As can be observed, the rotary ultrasonic rotary sonotrode's vibration amplitude grows linearly with the increase in the declared power of the ultrasonic generator.

The breaking forces of ultrasonic welds were then examined in relation to the stipulated power and welding speed using 42 groups of test specimens. Seven different welding speeds (0.077 m/s, 0.097 m/s, 0.113 m/s, 0.125 m/s, 0.147 m/s, 0.167 m/s, 0.227 m/s) were employed in six groups of declared ultrasonic generator powers. The ultrasonic weld breaking force measuring equipment, shown in Figure 3, was used to test the test specimens' breaking strength. The dimensions of the test specimens were ascertained using its test clamps, Figure 5b. Table 1 shows the outcomes of calculating the breaking forces. The designations of visual inspections of the ultrasonic welds before, Figure 7, and after tearing on the strength tester are included in addition to these data. Based on the collected data, functional relationships between the breaking forces of ultrasonic welds and the declared electric power of the ultrasonic generator and the speed at which the specimens were welded using the ultrasonic rotary sonotrode were formed (Figure 8). The maximum breaking forces for various ultrasonic generator powers and welding speeds are displayed very clearly in Figure 8. From the diagram, it can be deduced that an increase in the ultrasonic energy injected causes greater melting of the material and a higher breaking force of the weld. The breaking force of ultrasonic welds initially increases with an increase in the declared power of the ultrasonic generator. The breaking force can be increased until it reaches its limit, at which point the injected energy becomes excessive, causing excessive melting of the polymer material, harm to the weld, as well as a sharp decrease in breaking force. The plot in Figure 8 shows that the maximum breaking force for the utilized polymer material ranges from 32 to 34 N for various declared powers of the ultrasonic generator and welding speed. The largest breaking forces are attained at the four observed maxima in Figure 8, which correspond to the welding parameters' ideal values. These maximum values served as a check on the mathematical model's precision in accordance with expressions (1) to (32). The highest values were found for the welding speed of 0.077 m/s at the 200 W maximum power of the ultrasonic generator, 0.097 m/s at 250 W, 0.113 m/s at 268 W, and 0.125 m/s at 272 W.

It is necessary to calculate the amplitude of the oscillations of the rotary ultrasonic sonotrode using the calibrated diagram in Figure 6 or a mathematical expression that links the amount of oscillation amplitude to the declared power of the ultrasonic generator $A_0 = 0.484 + 0.3383 \cdot Pd$ for each of the four determined declared powers of the ultrasonic generator at which the maximum breaking force occurs. The sonotrode's real vibration amplitude can be found in values of 68.1, 85.1, 91.1, and 92.5 μm in Table 2 for the powers at which the highest breaking force occurs (200 W, 250 W, 268 W, and 272 W). The mathematical model's expression (28) can then be used to determine the ultrasonic sonotrode's effective power, $P\alpha$ (Table 2). The calculated effective power of the sonotrode is significantly lower than the declared power when the declared power of the ultrasonic generator and the effective power of the ultrasonic rotary sonotrode are compared. This discrepancy results from the utilization coefficient of the ultrasonic circuit (eclectic losses of the piezoelectric plates and other mechanical losses).

Using the mathematical model and the declared welding power, the necessary welding periods were computed (tm and ts). Table 2 shows the small differences that were calculated between the welding times actually accomplished and the required welding durations; these deviations are fairly acceptable for practical application. The heating of the weld's starting point can be attributed to the mathematical model's prediction of the delay time at which welding will occur tdy.

This can be used to demonstrate that, despite certain small simplifications, the mathematical model constructed is sound. Future scientific and applied research on ultrasonic welding of polymer materials can be well-founded on the mathematical model developed, in which 42 different factors are coupled by established mathematical interdependencies.

6. Conclusions

This paper discusses the potential application of a novel measurement technique. The complicated interconnectedness of the technological and technical parameters of ultrasonic welding of polymer materials is presented for the first time and discussed in this study. New accomplishments and novelty have also been made, which are reflected in the following ways:

1. A list of 44 characteristics has been put together and separated into three subgroups that are crucial for ultrasonic welding of polymer materials. There are 12 parameters in the first subgroup that are dependent on the polymer material being ultrasonically welded, 11 general acoustic and electroacoustic characteristics in the second subgroup, and 21 technical parameters in the third subgroup that are dependent on the ultrasonic welding equipment. Future research on ultrasonic welding of polymer materials may benefit from the systematization and cataloguing of these characteristics.
2. The new creation of comprehensive a mathematical model and the associated drawings are provided, further demonstrating the reliance of numerous significant parameters.
3. A suitable original measuring method and apparatus are described for calculating the vibration amplitudes of an ultrasonic rotary sonotrode.
4. The laboratory determination of breaking forces of test specimens welded by the ultrasonic welding method for polymer materials of various characteristics is described, together with the necessary measuring equipment due to the verification of the mathematical model.
5. The creation of calibration diagrams of the relationship between the ultrasonic rotary sonotrode's vibration amplitude and the declared power of the ultrasonic generator is also explained in order to get pertinent and realistic data needed for the implementation of the mathematical model presented.
6. Breaking forces were examined for 42 sets of test specimens that were welded at various welding speeds and welding powers on the basis of which the welding times were determined for the compounds with the maximum braking force and compared with the calculated ultrasonic welding times according to the mathematic model.
7. The area of the best welding parameters also was identified, along with typical variations in the functional dependency of breaking forces on the ultrasonic energy injected into the weld. The diagrams show areas where too little ultrasonic energy was applied, resulting in low breaking forces, optimal areas with ultrasonic energy applied, resulting in maximum joint breaking forces, and areas with too much ultrasonic energy applied. This resulted in deterioration of the welds.
8. The new developed mathematical model and the experimental data showed good agreement, with differences that are not important for practical application.

More extensive experiments on the presented model will be conducted in further research and papers. The authors hope that the mathematical model, which, for the first time combines such a large number of parameters, will serve as a basis for new comprehensive research in the field of ultrasonic welding materials. Future research will focus on testing mergers of different materials and comparability with other welding methods.

Author Contributions: Conceptualization, D.R. and S.F.; methodology, D.R. and S.F.; software, S.F. and S.F.R.; validation, S.F.R. and Ž.K.; formal analysis, D.R., S.F.R. and Ž.K.; investigation, D.R. and S.F.R.; resources, D.R. and Ž.K.; data curation, Ž.K. and S.F.R.; writing—original draft preparation, D.R. and S.F.R.; writing—review and editing, S.F. and S.F.R.; visualization, S.F.R.; supervision, S.F.R. and S.F.; project administration, S.F.R.; funding acquisition, D.R. All authors have read and agreed to the published version of the manuscript.

Funding: This research was funded by the Croatian Science Foundation through the project "Development and thermal properties of intelligent clothing (ThermIC), grant number IP-2018-01-6363".

Institutional Review Board Statement: Not applicable.

Informed Consent Statement: Not applicable.

Data Availability Statement: Not applicable.

Conflicts of Interest: The authors declare no conflict of interest.

References

1. Bhudolia, S.K.; Gohel, G.; Leong, K.F.; Islam, A. Advances in Ultrasonic Welding of Thermoplastic Composites: A Review. *Materials* **2020**, *13*, 1284. [CrossRef] [PubMed]
2. Jone, I. Ultrasonic and dielectric welding of textiles. In *Joining Textiles—Principles and Applications*, 1st ed.; Jones, I., Stylios, G.K., Eds.; Woodhead Publishing: Cambridge, UK, 2013; pp. 374–398.
3. Kuo, C.-C.; Tsai, Q.-Z.; Li, D.-Y.; Lin, Y.-X.; Chen, W.-X. Optimization of Ultrasonic Welding Process Parameters to Enhance Weld Strength of 3C Power Cases Using a Design of Experiments Approach. *Polymers* **2022**, *14*, 2388. [CrossRef]
4. Kayar, M.; Mistik, S.I.; Inan, D. Analysing effect of the factors on ultrasonic seam tensile properties of nonwoven fabrics by Nested Anova Design. *Int. J. Cloth. Sci. Technol.* **2015**, *27*, 803–817. [CrossRef]
5. Gomer, A.; Zou, W.; Grigat, N.; Sackmann, J.; Schomburg, W.K. Fabrication of Fiber Reinforced Plastics by Ultrasonic Welding. *J. Compos. Sci.* **2018**, *2*, 56. [CrossRef]
6. Sancaktar, E. Polymer adhesion by ultrasonic welding. *J. Adhes. Sci. Technol.* **1999**, *13*, 179–201. [CrossRef]
7. Shi, W.; Little, T. Mechanisms of ultrasonic joining of textile materials. *Int. J. Cloth. Sci. Technol.* **2000**, *12*, 331–350. [CrossRef]
8. Kiss, Z.; Temesi, T.; Bitay, E.; Bárány, T.; Czigány, T. Ultrasonic welding of all-polypropylene composites. *J. Appl. Polym. Sci.* **2020**, *137*, 48799. [CrossRef]
9. Nguyen, T.-H.; Thanh, L.Q.; Loc, N.H.; Huu, M.N.; Van, A.N. Effects of Different Roller Profiles on the Microstructure and Peel Strength of the Ultrasonic Welding Joints of Nonwoven Fabrics. *Appl. Sci.* **2020**, *10*, 4101. [CrossRef]
10. Atalay, O.; Kalaoglu, F.; Bahadir, S.K. Development of textile-based transmission lines using conductive yarns and ultrasonic welding technology for e-textile applications. *J. Eng. Fibers Fabr.* **2019**, *14*, 1558925019856603. [CrossRef]
11. Makwana, A.M.; Patel, V.R. Experimental Investigation of Ultrasonic Welding on Thermoplastic Material—A Review. *Int. J. Eng. Res. Technol.* **2017**, *6*, 350–352.
12. Palardy, G.; Villegas, I.F. On the effect of flat energy directors thickness on heat generation during ultrasonic welding of thermoplastic composites. *Compos. Interfaces* **2017**, *24*, 203–214. [CrossRef]
13. Chinnadurai, T.; Prabaharan, N.; Saravanan, S.; Karthigai, P.; Pandiyan, P.; Alhelou, H.H. Prediction of Process Parameters of Ultrasonically Welded PC/ABS Material Using Soft-Computing Techniques. *IEEE Access* **2021**, *9*, 33849–33859. [CrossRef]
14. Petriceanu, S.; Rontescu, C.; Cicic, D.; Bogatu, A. Ultrasonic characterization of the PVC welded materials. In *IOP Conference Series: Materials Science and Engineering*; IOP Publishing: Bristol, UK, 2018; Volume 849, pp. 1–8.
15. Khmelev, V.N.; Abramov, A.D. Model of Process and Calculation of Energy for a Heat Generation of a Welded Joint at Ultrasonic Welding Polymeric Thermoplastic Materials. In Proceedings of the 8th Siberian Russian Workshop and Tutorial on Electron Devices and Materials, Novosibirsk, Russia, 1–5 July 2007.
16. Dieulesaint, E.; Royer, D. *Elastic Waves in Solids*, 1st ed.; John Wiley & Sons: Hoboken, NJ, USA, 1980; pp. 239–240.
17. Pfaff Industrial. PFAFF 8310 -041/002 with Sonotrode from Top, Feed-off-the-Arm Version (from Side). Available online: https://www.pfaff-industrial.com/en/portfolio/welding-machines/ultrasonic-sealing-machines/pfaff-8310-042-sontrode-unten (accessed on 7 August 2022).
18. InfinitiVision 3000A X-Series Oscilloscopes, Data Sheet. Available online: http://www.keysight.com/ (accessed on 7 August 2022).
19. Fiber Optic MTI 2100 Fotonic Sensor. Available online: https://mtiinstruments.com/products/non-contact-measurement/fiber-optic-sensors/fiber-optic-mti-2100-fotonic-sensor/ (accessed on 7 August 2022).
20. Silva Girão, P.; Postolache, O.; Faria, J.B.; Pereira, J. An Overview and a Contribution to the Optical Measurement of Linear Displacement. *Sens. J. IEEE* **2002**, *1*, 322–331. [CrossRef]
21. Sauter model HF 500, Scales and Measuring Instruments. Available online: https://scales-measuring.com/en/force-measurement/2172-digital-force-gauge-sauter-fh-500.html (accessed on 7 August 2022).
22. Sauter Electrical Horizontal Test Stand THM 500N500N. Available online: https://www.kern-sohn.com/shop/en/measuring-instruments/force-gauges/THM/ (accessed on 7 August 2022).

machines

MDPI

Article

Non-Destructive Disassembly of Interference Fit under Wear Conditions for Sustainable Remanufacturing

Ho Lam Au-Yeung [1], Sabbah Ataya [2], Hany Hassanin [3,*], Mahmoud Ahmed El-Sayed [4], Mahmoud Ahmadein [5], Naser A. Alsaleh [2], Mohamed M. Z. Ahmed [6] and Khamis Essa [1]

1. School of Engineering, University of Birmingham, Birmingham B15 2TT, UK
2. Department of Mechanical Engineering, Imam Mohammad Ibn Saud Islamic University (IMSIU), Riyadh 11432, Saudi Arabia; naalsaleh@imamu.edu.sa (N.A.A.)
3. School of Engineering, Technology, and Design, Canterbury Christ Church University, Canterbury CT1 1QU, UK
4. Department of Industrial and Management Engineering, Arab Academy for Science Technology and Maritime Transport, Alexandria 21599, Egypt; dr.mahmoudelsayed12@gmail.com
5. Department of Production Engineering and Mechanical Design, Tanta University, Tanta 31111, Egypt; m.ahmadein@f-eng.tanta.edu.eg
6. Mechanical Engineering Department, College of Engineering at Al Kharj, Prince Sattam Bin Abdulaziz University, Al Kharj 16273, Saudi Arabia; moh.ahmed@psau.edu.sa

* Correspondence: hany.hassanin@canterbury.ac.uk

Abstract: Remanufacturing has been gaining increasing attention in the last few years as a part of green engineering. It is the process of restoring the original specifications of a given product utilizing a combination of new, repaired, and old parts. The present study investigates non-destructive disassembly of an interference fit pin-hub joint to enable the reuse of worn parts with the same loading capacity. The aim is to reduce the disassembly force while preventing plastic deformation and frictional damage on the contact surface to avoid fretting failure and enable further coating. A finite element model of a shaft/hub interference fit was developed, taking into account two cases of damage to the mating parts: deformation and corrosion. The results indicate that thermal disassembly is effective in reducing breaking force by 50% in deformed joints, whereas vibration waves are more suitable for corroded parts with increased friction. In addition, applying a low-frequency oscillation force to the axis of disassembly reduces the pulling out force by 5% and plastic deformation by 99% due to acoustic softening effects. Furthermore, using a heat flux simultaneously with vibration decreases the breaking force by 85%, indicating the higher effectiveness of thermal-aided disassembly and vibration-assisted disassembly in reducing the breaking force of corroded parts with increased friction. This study provides remanufacturing designers with efficient tools to weaken the interference fit and decrease the disconnecting force, ultimately reducing the cost and time required for the disassembly process.

Keywords: interference fit; press-fit pin-hub; automated disassembly; remanufacturing; finite element analysis

1. Introduction

In recent years, there has been growing interest in remanufacturing as a research topic, especially in the context of sustainability and green engineering. Remanufacturing is the process of restoring used products to a condition that is equal to or better than their original state, thereby minimizing waste and reducing carbon footprints. The process involves various steps, such as sorting, disassembly, cleaning, inspection, and rebuilding [1,2]. Disassembly of fitted components, particularly interference-fitted ones, is a critical step in the remanufacturing process. Disassembly is essential for product recovery and is the primary source of data related to remanufacturing operations. Interference fits, also referred to as friction fits or press-fits, are commonly employed in mechanical joints due to their cost-effectiveness and simple design. These joints are particularly useful for cylindrical components in various mechanical systems because of their precise radial alignment [3]. Torque-transmitting bearings, which are widely used in heavy machinery, locomotives, and electrical hardware systems [4], are some of the applications of press-fit joints. Disassembly of interference joints is a crucial stage in the disassembly process. However, due to the complexity of the mating parts, it can be challenging to remove them without causing plastic deformation, scratches, and adhesion. These types of friction damage can adversely affect the success of the remanufacturing process.

Remanufacturing, a process aimed at maintaining the original shape and/or output of a component, is in contrast to recycling. Successful remanufacturing depends on the integrity of the parts during disassembly, which is a crucial step. Unfortunately, disassembly is mostly performed manually with hand tools, as automated tools are expensive [5], leading to increased labor costs and limiting the applicability for complex products [6]. For example, disassembling shaft-and-hub joints often requires applying torque and axial loading, which can cause deformations and increase separation difficulty. Additionally, the loss of contact between the shaft and hub increases susceptibility to fretting wear [7]. Earlier research by Radi et al. developed a mathematical model validated by finite element analysis (FEA) to investigate the bending moment of the detachment of shaft-hub press-fits [8]. Hammond et al. demonstrated in another study that disassembly was a significant concern for remanufacturers, with corrosion being the most serious concern during disassembly [9]. Corrosion can cause sticking or bonding of the parts, making disassembly difficult, particularly after prolonged storage of up to 30 years, which can initiate adhesive and corrosion bonds. This increases the extraction force needed, as the coefficient of friction (COF) increases [10]. The required disassembly power for rusted joints could be twice that of non-corroded ones [11].

Permanent fastening is considered a significant challenge during disassembly processes, second only to corrosion. Breaking up interference fits can be a difficult process that often results in frictional damage and plastic deformation. The amount of force required for removal is directly related to the extent of the damage caused [2]. To overcome this problem, non-destructive disassembly (ND) methods have been proposed to prevent damage to components [9]. A study by Mok et al. showed that minimal extraction force was one of the most critical characteristics for easy disassembly of mechanical parts [12]. Reducing the extraction force is crucial as it allows the application of simple tools and techniques and minimizes frictional damage, thereby reducing refurbishing costs. However, interference fits, also known as permanent or semi-permanent fittings, are often considered impossible-to-disassemble [13]. When conducting non-destructive disassembly (ND), it is crucial to maintain the loading capacity of interference fits while avoiding surface damage, such as bulges or furrows. Disassembly should be carried out in a manner that avoids plastic and thermal strains, as plastic strain can result in permanent deformation of the structure, which can ultimately affect the strength and ductility of the part [14]. When a part is subjected to plastic strain, it becomes more susceptible to necking, which significantly reduces the life cycle of the remanufactured part [7]. Zhou et al. [2] analyzed plastic strains and found a strong correlation between surface damages and extraction force. Consequently, significant efforts have been made to reduce extraction forces during disassembly processes.

Currently, various methods are used to ease the extraction of interference fits, such as cold contraction, thermal expansion, and injection of pressurized oil [14]. The primary goal of these methods is to minimize contact pressure and reduce the force required for disassembly. Two primary approaches have been studied: thermal disassembly and vibration-assisted disassembly. Thermal disassembly has several advantages, including reducing surface damage, being highly effective for coatings in post-processing and refurbishing, and being suitable for automated disassembly. Wang et al. [15] explored the use of liquid nitrogen to weaken a sleeved interference fit. The sleeved joints allowed the liquid nitrogen to pass through and create radial shrinkage. Their findings showed that this technique significantly reduced the radial and hoop stresses, which, in turn, reduced the disassembly force required. However, the study did not investigate thermal strains and plastic deformation.

Shaheen et al. [16] conducted a study to investigate the thermal behavior of high-strength stainless steel bolt assemblies at elevated temperatures. They observed that the tensile strength of the assemblies gradually decreased with increasing temperature up to 300 °C, followed by a much sharper drop. Beyond 300 °C, the assemblies could not retain much of their original strength even after cooling. It is crucial to maintain the original strength of the parts after thermal expansion to ensure acceptable performance after remanufacturing. Sun et al. [17] also studied the behavior of S235 steel under an elevated temperature and suggested that thermal plastic strains significantly increased at temperatures above 200 °C. In another study, Bengeri et al. [18] investigated the thermal assembly of a shrink fit and concluded that the residual stresses, particularly the interface pressures, were not significantly affected by the drop in yield stress at high temperatures, and hence could be considered approximately isothermal. However, this study cannot be directly applied to thermal disassembly as it did not take into consideration the contact area between the shaft and hub prior to assembly, where pretension and pressure at the contact surfaces had not yet occurred. The changes in pressure might be weakened due to the heat transfer between the contact zones.

By contrast, vibration-assisted disassembly, specifically using ultrasonic micro-vibrations, has been utilized to aid in the disassembly of press-fit assemblies by employing low-frequency oscillations. These vibration waves can decrease friction and extend the elastic–plastic range, potentially avoiding damage during the disassembly process [15]. Studies have demonstrated that vibration waves can reduce the force required to break up an interference fit by up to 60% while preserving the properties of the parts. Furthermore, low-frequency oscillation waves can improve and refine the surface texture by altering the microstructure and removing weaker dirt, which can be further enhanced by combining vibration with chemical fluids and abrasive tools, ultimately reducing the time and effort necessary for refurbishing parts after disassembly [19]. Thus, this type of wave is well-suited for corroded surfaces with a higher coefficient of friction and loss of materials. Fridman et al. observed a significant decrease in friction between sliding surfaces as a result of the application of vibration, resulting in a significant weakening of the connection between assembled parts [20]. The impact of vibration waves is highly dependent on tangential stiffness, stresses, and surface roughness, as reported by Gutowski et al. [21]. Mikolainis and Bakšys used an electromagnetic vibrator to apply vibration waves to an interference-fit assembly and found that the force and duration required for disassembly could be reduced by adjusting oscillation parameters, namely the amplitude (A) and frequency (f) of vibration [22]. The exposure and duration of oscillation waves were also found to impact the weakening of connections. Several researchers have demonstrated an inverse effect of ultrasonic vibration on the force required for disassembly [23]. The three widely recognized explanations for the influence of ultrasonic vibration on metal plasticity are stress superposition, acoustic softening, and reduced friction. Nonetheless, the underlying mechanism responsible for this phenomenon is still unclear and remains a topic of debate to date [24].

While there has been extensive research on interference fit disassembly, particularly in ideal conditions, limited attention has been given to the disassembly force of an interference fit under normal wear conditions. The presence of corrosion, bonding, and deformation between the mating parts can significantly increase the disassembly force required. The identified knowledge gap served as the motivation for the current study, in which a finite element analysis was utilized to model a press-fit shaft and hub. The study aimed to investigate potential methods for facilitating disassembly during remanufacturing, with a focus on friction as a significant factor influencing the disassembly force. Our approach involved comparing the forces required for disassembly before and after applying thermal expansion and vibration techniques, with a focus on examining plastic strains. This study adds a novel contribution to the existing literature on interference fit disassembly by exploring ways to address the practical challenges faced in real-world disassembly scenarios.

2. Methodology

This study focused on investigating two common problems that arise during the disassembly process: deformation and corrosion of the mating components. To address these issues, thermal expansion and vibration-assisted disassembly techniques were examined. The research team developed a baseline model of an interference fit and validated it using literature data. The model accounted for normal wear of the joint and considered two scenarios: deformation caused by bending of the connected parts and an increase in COF due to corrosion. The study explored various approaches to reduce the pull-out force, minimize plastic strain in the parts, and prevent damage during and after disassembly. Figure 1 provides a block diagram of the modelling sequence used in this study.

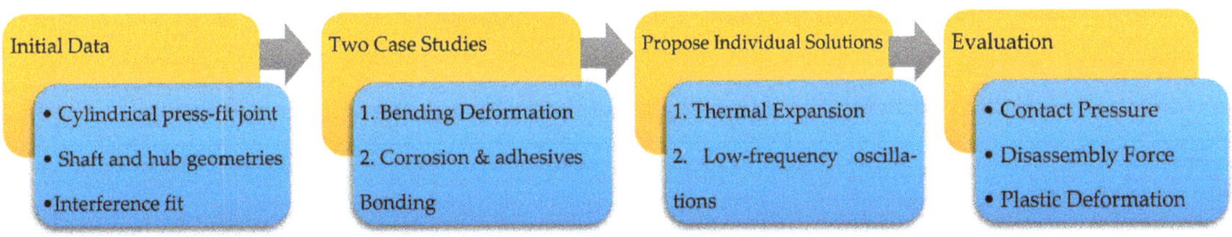

Figure 1. Workflow of the current study.

This study used a finite element model developed by Madej et al. [25], which simulated the press-fit process of a simple pin-tube mechanical joint with different tolerances between the mating shaft and hole. Madej et al. analyzed both pressing and withdrawal modes and validated the model using experimental data from the same study and by comparing the assembly force graphs with those from another study conducted by Croccolo et al. [26]. Based on the similar trends observed in the results, this model was selected as the baseline model for the current study.

The finite element model for the press-fit joint was created using Ansys 2021, with the dimensions and shape of the shaft/hub fit shown in Figure 2a. This fit corresponds to the J10 (O = 0.022 mm) specification as per the reference literature [27]. The joint in this study follows the H7/s6 standard, which is a medium drive fit used for permanent assembly, commonly found in drive gears and sleeve bearings. Due to the difficulty in disassembling this type of joint without causing damage, it was selected for the current study. The material chosen for this study was S235 structural steel, which is a low-carbon manganese steel. S235 steel was selected for its common use in manufacturing, particularly in the automotive industry where remanufacturing is important. It has good weldability and machinability and is relatively inexpensive, making it a cost-effective choice for remanufacturing processes. Using this material allowed for a realistic and relevant analysis of the challenges associated with disassembling parts for remanufacturing.

To model the pressing assembly process, finite element analysis (FEA) was utilized as depicted in Figure 3, replicating the model developed by Madej et al. [25]. The shaft was displaced by 15 mm in the Z-axis direction, with a step size of 0.5 mm. The material properties and plasticity data of the S235 steel alloy presented in Tables 1 and 2 were applied to the mating parts (shaft and hub) in the model. The contact surface was modelled using the augmented Lagrange function with a COF of $\mu = 0.1$. A support was fixed at the front outer surface of the hub, as shown in green (see Figure 2b), following the method described in the literature [25]. The maximum force over time was considered as the force required to assemble the joint, whereas the force required to displace the shaft from the hub by 0.1 mm was taken as the breaking force value. The baseline model was validated by comparing the obtained results with those from the literature, as described below. Subsequently, it was used to further analyze the interference fit disassembly. Section 2.1 considered a bending couple to study deformed interference fits, while Section 2.2 investigated corrosion, which is classified as the most significant problem in ND.

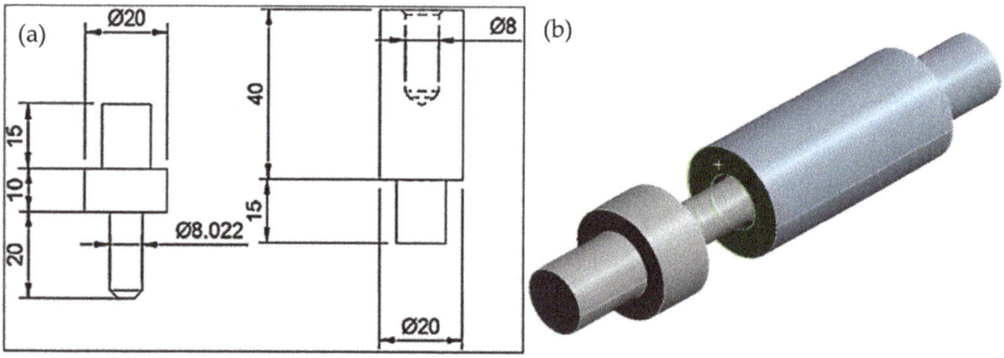

Figure 2. (**a**) A sketch of the shaft/hub interference fit and (**b**) the CAD model. Dimensions are in mm.

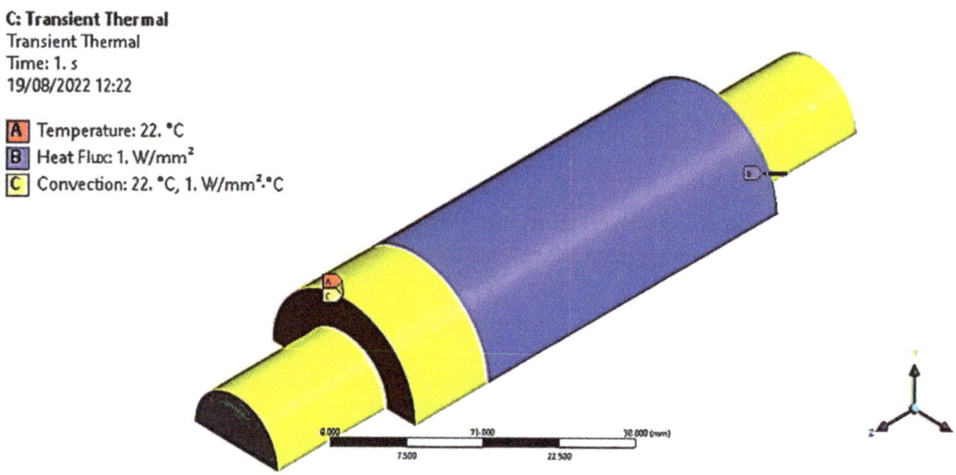

Figure 3. Case 1 transient thermal model setup.

Table 1. Chemical composition of S235 alloy.

Element	C	Mn	P	S	N	Cu	Fe
%	0.17	1.4	0.035	0.035	0.12	0.55	Bal.

Table 2. S235 steel alloy mechanical properties.

Young's Modulus (GPa)	200
Yield Strength (MPa)	235
Ultimate Tensile Strength (MPa)	360

2.1. *Thermal Expansion-Assisted Disassembly*

Interference fits can lead to bending and torsion of the mating components, which increases their susceptibility to fretting and fatigue [13]. Peaking stress occurs when the shaft and hub are subjected to bending loads, resulting in relative motion between them. Torsion only increases shear motion but may cause plastic deformation under high-stress conditions when slipping is prevented. According to Radi et al. [8], the mating parts experience increasing pressure at the concave side and vice versa. The detachment bending couple is the moment needed to detach a hub from a shaft due to frictional forces at the interface. A mathematical model has been derived to calculate the detachment couple C of a shaft-hub interface fit as shown in Equation (1), where E is the Young's modulus, I is the diametrical interference, I_1 is the moment of inertia, and a is the shaft radius. The detachment couple is then applied to the shaft, as shown in Figure 3, while the other end of the hub is fixed. The primary objective of this study is to prevent the increase in plastic deformation at the shaft/hub interface during disassembly, which is crucial for ND.

$$C \approx \frac{0.4IEI_1}{a^2} = 0.32IEa^2 \quad (1)$$

To facilitate disassembly through thermal expansion and reduce contact stress, an even heating process was applied to the hub. This approach is similar to shrink fitting, where a transient heat is applied to the hub during assembly, causing it to shrink while cooling after assembly [28]. Since the hub has a significantly larger surface area than the pin, the expansion of the hub would have a greater effect. To simulate actual heat transfer during disassembly, a transient thermal analysis was performed using Ansys Workbench 2021, as illustrated in Figure 4. A heat flux was applied to the exterior of the hub, and both ends were assumed to be at room temperature, with convection taken into account. The initial temperature was maintained at 22 °C, and a displacement of 4.5 mm was applied to the shaft. The Ansys model implemented the mechanical and thermal properties of S235 steel at different temperatures, obtained from the literature [29], and these are presented in Figure 4. The study investigated the effect of thermal heating within a temperature range of less than 300 °C, as both the yield strength and Young's modulus of the alloy decreased with increasing temperature. The elasticity was determined by Young's modulus (E) and the tangent modulus, with the model keeping the latter constant. Young's modulus is a measure of a material's stiffness under axial tension or compression and is calculated as the ratio of stress to strain within the proportional limit. It is a constant value for a given material and is independent of the magnitude of the stress applied. By contrast, the tangent modulus is a measure of the local stiffness of a material at a specific point on its stress–strain curve and is calculated as the slope of the curve at that point. Unlike Young's modulus, the tangent modulus is not a constant value but varies depending on the level of stress applied. It is worth noting from Figure 4 that the thermal conductivity decreased by more than 13% after 300 °C, leading to less efficient heat transfer. Additionally, the tensile strength also experienced a drastic decline after 300 °C. Therefore, a heat flux in the range of 0.5 to 2 W/mm^2 was applied to avoid violating this constraint and keeping the part's temperature below 300 °C, as shown in Figure 4. The study determined the plastic strain, contact pressure, and disassembly force resulting from the transient thermal analysis.

2.2. Vibration-Assisted Disassembly

The S235 alloy, being a mild steel, is vulnerable to uniform corrosion when exposed to humid air. Previous studies by Gassama et al. and Kocanda et al. have shown that the coefficient of friction (COF) increases significantly when wear occurs in seawater [31,32]. These studies assumed that no anti-corrosion coatings were applied to the contact surfaces of the joint. In line with these findings, a COF of 0.6 was used in the current study for ANSYS static structural analysis. The selection of the coefficient of friction (COF) value of 0.6 was carefully considered and based on a thorough literature review and common industrial practices for similar applications. This value falls within the typical range for metal-on-metal contacts, such as those found in automotive components, and has been widely used in simulations and experiments related to remanufacturing processes. The wear effect was considered a combination of mechanical, abrasive, and corrosive factors, while the effects of time-dependent corrosion and loss of material were neglected. Previous studies by Dieudonné et al. and Gutowski et al. investigated the impact of longitudinal vibrations on friction forces in sliding motions [15,21]. Both studies applied oscillations of different frequencies to the support to simulate the acceleration force exerted on the hub in the same axis of sliding motion (i.e., the z-axis in this FEM). In this study, vibration waves were applied as sinusoidal forces described by two functions presented in Equations (2) and (3). The force F varied with time t, and two different vibration amplitudes (1000 N and 5000 N) were considered while the frequency remained constant.

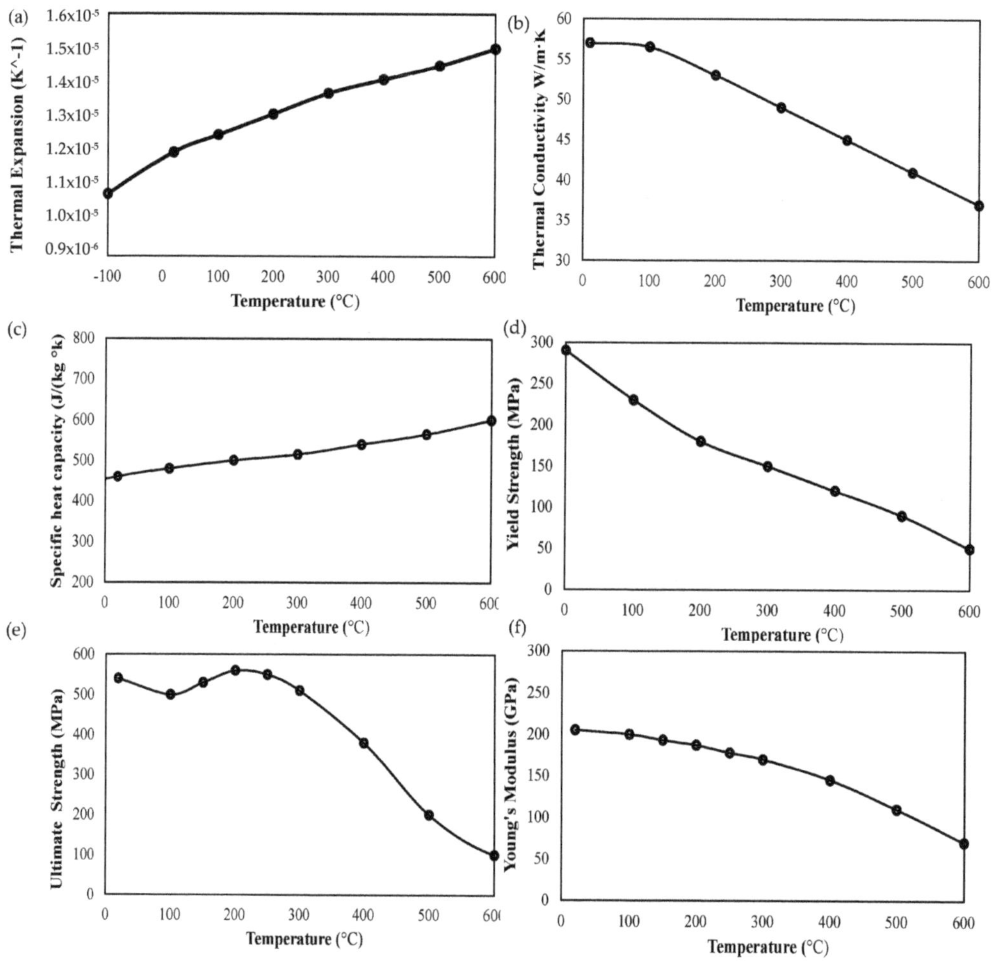

Figure 4. Physical and mechanical properties of S235 alloy vs. temperature (**a**) The coefficient of thermal expansion, (**b**) thermal conductivity, (**c**) specific heat capacity, ref. [30] (**d**) yield strength, (**e**) tensile strength, and (**f**) Young's modulus. Reprinted/adapted with permission from [18] 2023, Elsevier.

$$F = -A\, sin^2\left(\frac{2\pi t}{0.0002}\right) f \approx 700 \text{ Hz} \quad (2)$$

$$F = -A\, sin^2\left(\frac{2\pi t}{0.0007}\right) f \approx 1500 \text{ Hz} \quad (3)$$

To examine the impact of higher frequencies on the disassembly process, a threshold of 1500 Hz was selected as a relatively high frequency [33]. This was performed to bridge the gap between previous research conducted by Gutowski et al. [21], who studied high frequency (4000 Hz), and Dieudonne' et al. [3], who examined low frequency (4–60 Hz). To further explore the impact of lower frequency, a frequency of 700 Hz was also included. During the disassembly process, vibration waves were applied simultaneously, while an extraction of 2 mm with a step of 0.2 mm was carried out. The breaking force was determined as the total force required to move 0.1 mm of the shaft. All other settings remained consistent with the baseline model described earlier.

3. Results and Discussion

In this section, the results of the study on the disassembly force using developed FEA model are presented. The accuracy of the model is evaluated by comparing its results to those reported in the literature. The disassembly force predicted by the model and those reported in the literature [26] are compared in Figure 5a. The model estimates the breaking and assembly forces to be 8986 N and 6002 N, respectively, which is slightly higher than the literature-reported values of 8590 N and 5469 N. The model's ability to accurately predict the disassembly force under different conditions indicates its usefulness in optimizing the design of bolted joints for specific applications. Additionally, the model predicts a maximum surface pressure of 267 MPa, as shown in Figure 5b, which is slightly lower than the literature-reported value of 277 MPa. The difference may be due to the assumptions made in the model, such as the assumption of a perfectly flat contact surface between the bolt and the nut, which may not hold in real-world scenarios. However, the margin of error of approximately 5% indicates an acceptable level of accuracy in predicting the disassembly force under different conditions in the current study. Figure 5b also shows that the inlet experiences a significant stress concentration, consistent with the findings of Pedersen [13]. This stress concentration may be attributed to the geometric features of the inlet, such as its sharp edges and corners, which can act as stress concentrators. The identification of these stress concentrations can aid in the optimization of the bolted joint design to prevent failure. As discussed, the predicted assembly force was initially higher than that in the literature [25]. This discrepancy may be due to the relatively larger overlapping areas of the shaft and hub in the current study. The model assumed an overlapping of 0.1 mm to set up the contact area roughness, which may have resulted in a higher predicted assembly force. However, this difference is not significant for the current study, as the primary objective was to reduce the breaking forces of the joint.

To further validate the results of the FEA model, a convergence analysis of the maximum equivalent stress was conducted, as shown in Figure 5c. A mesh element size of 1 mm was selected, as the difference in the results was within 5%. This mesh element size was applied to the two cases considered in this study.

3.1. Thermal-Assisted Disassembly

The results of the study on the decoupling process of a shaft-hub joint using numerical simulation are presented in this section. Equation (1) was used to determine the decouple moment, which was found to be approximately 25 kN mm. This value was then applied to the shaft, as shown in Figure 6a. It is worth noting that the model used in this study had a protruding shaft. Therefore, the shaft-hub contact force was not directly loaded, and the shaft was only deformed compared to non-protruded ones [8]. Figure 6b shows that plastic strain had developed around the edges of both the shaft and the hub. The goal of the numerical decoupling (ND) process was to avoid increasing such deformation and maintain the plastic strain as before disassembly.

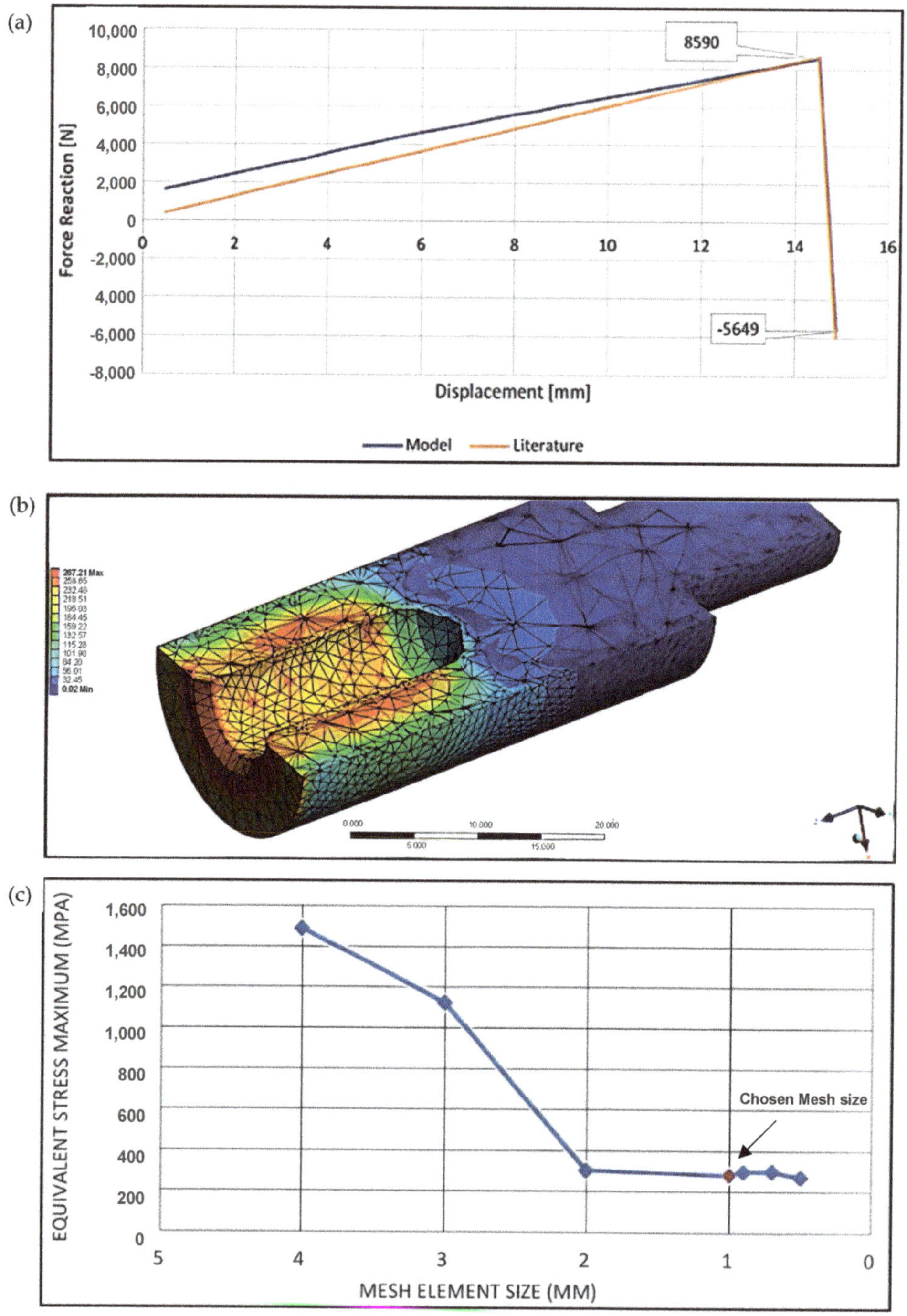

Figure 5. (**a**) Assembly force with marked breaking force comparisons [25], (**b**) surface pressure after assembly, and (**c**) mesh convergence graph.

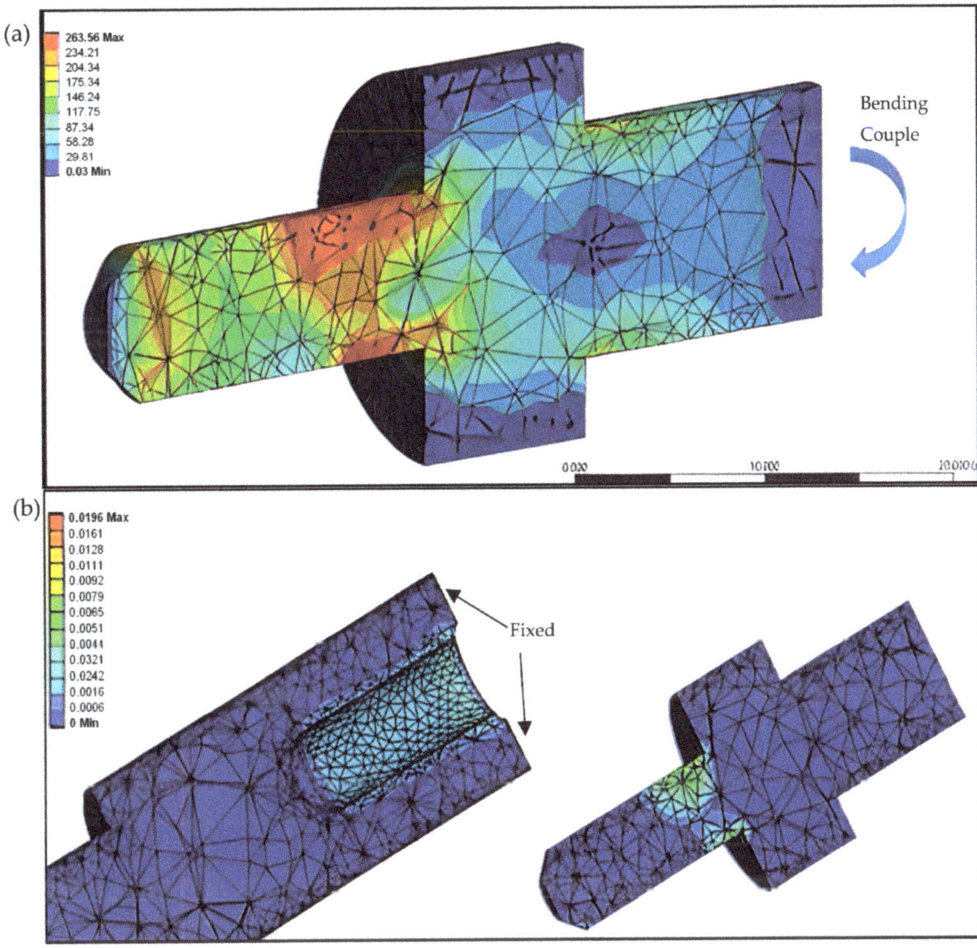

Figure 6. (**a**) Maximum stress at a moment of 25 kN mm and (**b**) plastic strain after deformation in the shaft and the hub.

Figure 6b illustrates that the plastic strain was confined to the edges of the shaft and hub, indicating that the elastic deformation was mainly responsible for transmitting the load. The model predicted that the maximum plastic strain was around 0.019, which is within the acceptable range for many engineering applications. The results of this study demonstrate that ND can effectively reduce the deformation and maintain the plastic strain of a shaft-hub joint during disassembly. This approach can potentially reduce the risk of joint failure and increase the lifespan of the joint. The model's ability to accurately predict the plastic strain and deformation can help optimize the design of shaft-hub joints for specific applications.

The change in contact pressure with increasing applied heat flux is analyzed using Figure 7a. The maximum and average pressures are observed to decrease significantly as the heat flux is increased. Initially, with no heat applied, the maximum and average pressures are approximately 680 MPa and 230 MPa, respectively. However, at a heat flux of 2 W/mm^2, these values drop to approximately 250 MPa and 30 MPa, respectively. These results are consistent with those reported by Sen et al. [29], who observed similar trends in both radial and hoop stress under the application of thermal energy. The reduction in contact pressure can be attributed to the thermal expansion of the hub and shaft, which reduces the compressive forces between the two components. Additionally, the local increase in temperature causes a decrease in the modulus of elasticity and an increase in the coefficient of thermal expansion, which further contributes to the reduction in contact pressure. It is important to note that the reduction in contact pressure can have significant implications for the structural integrity of the joint. In particular, it can increase the risk of fatigue failure and reduce the lifespan of the joint. Therefore, it is essential to carefully consider the effects of thermal loading when designing and analyzing mechanical joints.

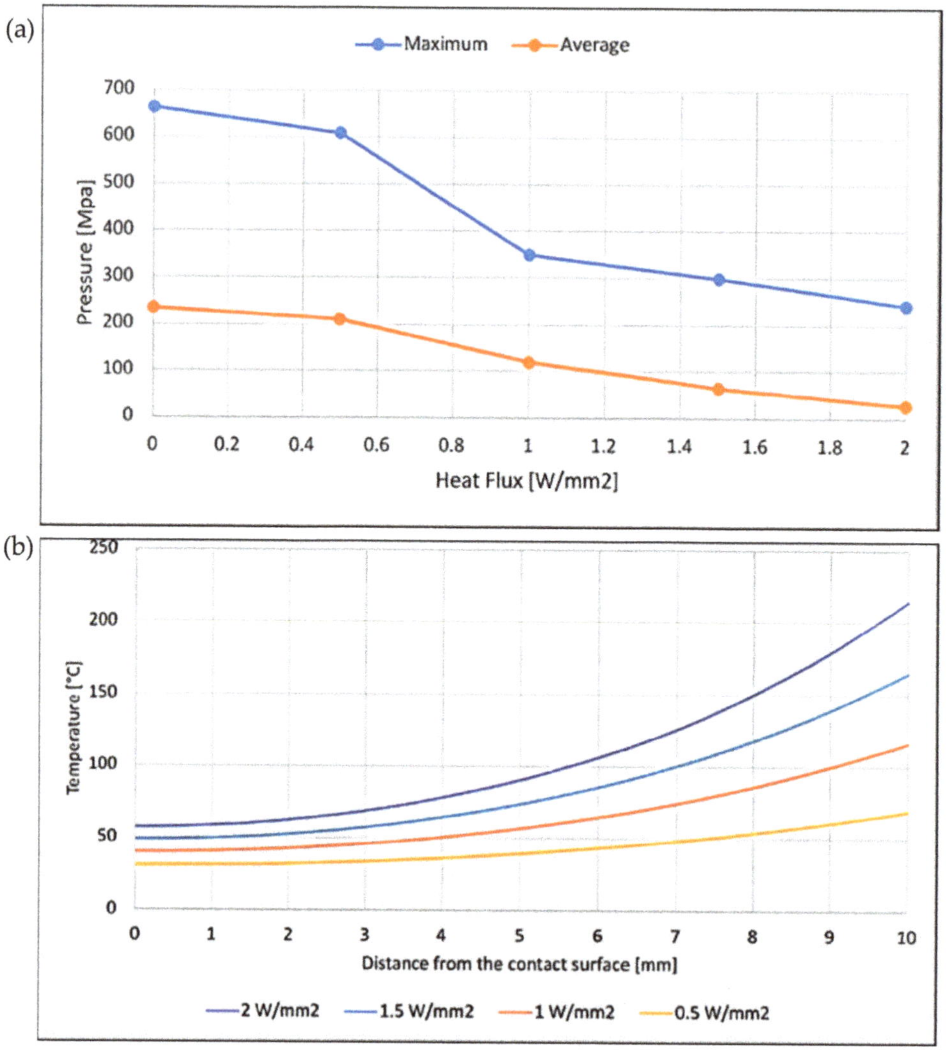

Figure 7. (**a**) Change in contact pressure with heat flux and (**b**) temperature change along radial distance from shaft surface to hub exterior at different heat fluxes.

The results showed that the applied heat flux had a significant effect on the temperature, contact pressure, and disassembly force of interference shaft-hub connections. As shown in Figure 7b, the temperature along the external surface increased consistently with an increase in the stance from the contact surface, with the highest occurring at a heat input of (1 W/mm^2).

The temperature rise shown in Figure 7b was found to be associated with a reduction in the disassembly force. Figure 8a shows the lowest force required for disassembly was shown the least for the highest heat input (1 W/mm^2). Figure 8b illustrates the temperature gradient through the joint at an applied heat flux of 2 W/mm^2. Notably, the effect of the applied heat flux is reduced at the interface between the shaft and hub, compared to the external heating surface. This is because the Z-axis contact line, which is the most important zone affecting heat transfer in an interference shaft-hub connection, experiences a reduction in contact pressure due to the thermal expansion of the hub and shaft.

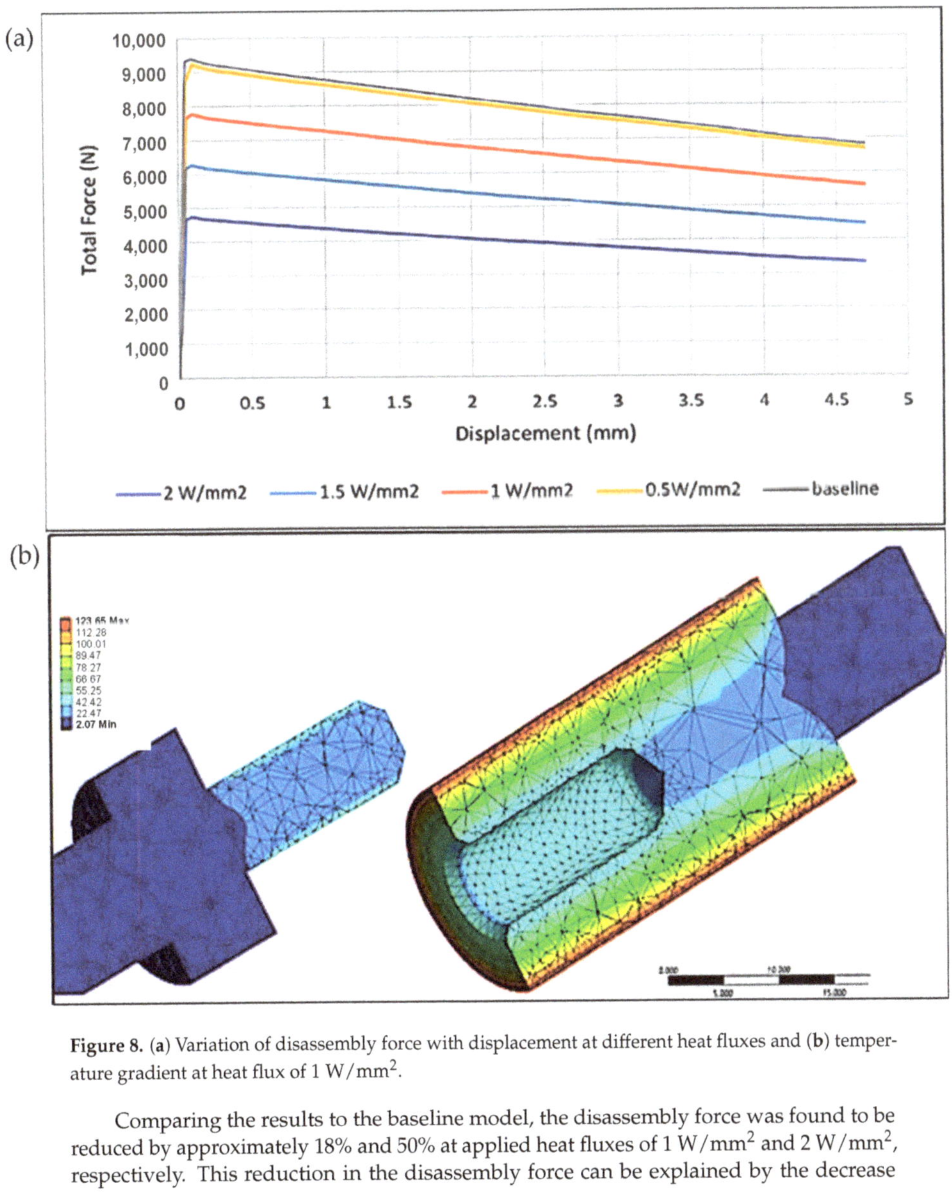

Figure 8. (**a**) Variation of disassembly force with displacement at different heat fluxes and (**b**) temperature gradient at heat flux of 1 W/mm^2.

Comparing the results to the baseline model, the disassembly force was found to be reduced by approximately 18% and 50% at applied heat fluxes of 1 W/mm^2 and 2 W/mm^2, respectively. This reduction in the disassembly force can be explained by the decrease in contact pressure along the Z-axis contact line, which is consistent with the findings of Sen et al. [29]. These results have practical implications for the design and analysis of interference shaft-hub connections subjected to thermal loading, particularly in high-temperature applications. The reduction in the disassembly force observed in this study suggests that applying heat to the connection before disassembly could be an effective means of reducing the force required for disassembly. Additionally, the results highlight the importance of considering the effect of thermal loading on interference shaft-hub connections in the design and analysis process.

The values (both maximum and average) of plastic and thermal strains after disassembly, at different amounts of the heat flux applied, are given in Table 3. The application of heat flux resulted in a reduction of approximately 7% and 3% in average plastic strain for heat fluxes of 0.5 and 1 W/mm^2, respectively, compared to direct disassembly (without heating). However, the maximum plastic strain remained constant with an increasing heat flux up to 1 W/mm^2 and then experienced a significant surge of about 140% when the heat flux was increased to 2 W/mm^2. This increase in the maximum plastic strain can be attributed to the induced thermal strain. The observed thermal strain increased by approximately 2500% as the heat flux was increased from 0.5 to 2 W/mm^2, as shown in Table 3. These results are consistent with previous experimental findings by Sun et al. [17], who reported a substantial rise in thermal plastic strains with increasing temperature above 200 °C for alloy S235. The study also found that temperature effects may induce solid plasticity and mechanical loads [34]. As the ultimate tensile strength and the Young's modulus of the alloy decreased after reaching 200 °C, as shown in Figure 4, strains developed at lower stress values, which could explain the increase in the overall average plastic deformation to reach 1.73×10^{-4}. It should be noted that the current work did not study the cooling after thermal disassembly and the thermal strain-hardening effect. Further experimental studies would be required to allow more accurate determination of the actual plastic deformations and disassembly forces considering such influencers.

Table 3. Plastic and thermal strains after disassembly at different heat fluxes.

Heat Flux (W/mm^2)	Maximum Plastic Strain	Average Plastic Strain	Maximum Thermal Strain	Average Thermal Strain
0	4.21×10^{-3}	1.42×10^{-4}	-	-
0.5	4.21×10^{-3}	1.32×10^{-4}	3.60×10^{-5}	2.47×10^{-5}
1	4.21×10^{-3}	1.38×10^{-4}	3.36×10^{-4}	2.31×10^{-4}
2	1.00×10^{-2}	1.37×10^{-4}	9.35×10^{-4}	6.43×10^{-4}

3.2. Vibration-Assisted Disassembly

Figure 9 displays the baseline disassembly force versus displacement for both the deformation and corrosion cases. The vibration waves were applied as sinusoidal forces using Equations (2) and (3) as described earlier. From the figure, it is evident that increasing the coefficient of friction (COF) to 0.6 resulted in a significant increase in the required breaking force. Specifically, the breaking force increased by a factor of about 2.5, which indicates an increase in the difficulty of disassembly. The use of vibration waves for disassembling interference-fit assemblies can be further considered for future studies. In particular, the contact response with cyclic micro-slips and frictional dissipation in the joint to fully understand its feasibility and safety. The force amplitudes presented only refer to excitation. Finite element analysis with a contact algorithm can simulate the interaction between the shaft and hub, including friction and micro-slips, to determine the potential for joint damage or failure during disassembly.

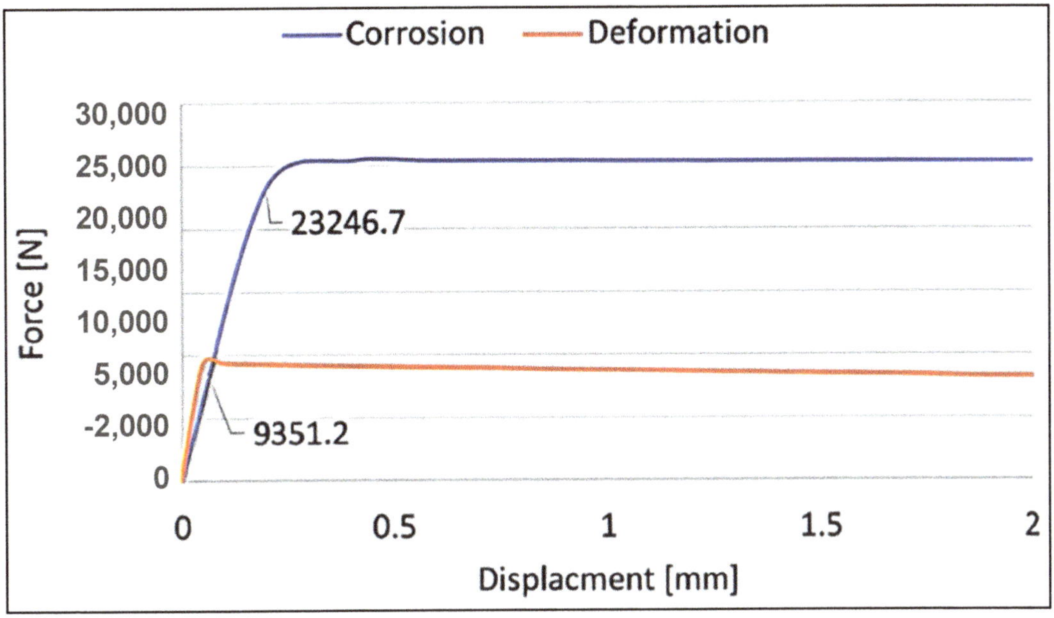

Figure 9. Baseline disassembly force for the deformed and corroded joints; labelled breaking force at 0.1 mm displacement.

In this study, the motion graphs of different force amplitudes at frequencies of 700 and 1500 Hz were analyzed and presented in Figure 10a,b, respectively. The harmonic response analysis was performed using Ansys to determine the natural frequency of the joint. It was found that the natural frequency of the joint was significantly higher than that of the applied vibrations. Therefore, the joint was not expected to achieve resonance, which could cause irreversible damage. This indicates the feasibility and safety of the proposed method for the disassembly of interference-fit assemblies through vibration waves.

Figure 11 presents the effect of oscillation waves with various amplitudes and frequencies, as well as an increase in joint temperature, on the disassembly force. The results related to the thermal effect are plotted with respect to a secondary axis to the right. The baseline model required a disassembly force of approximately 25.4 kN. The application of a vibration wave with an amplitude of 1000 N and a frequency of 700 Hz did not have a significant effect. However, when either the amplitude or frequency was increased, the driving force was shown to decrease, particularly with increased displacement. For example, when a vibration wave with an amplitude of 5000 N and a frequency of 1500 Hz was applied, the breaking force at a displacement of 2 mm was about 24.6 kN. These findings are consistent with previous experimental studies that suggest that the driving force level can be controlled by changing the vibration amplitude or frequency [3,21]. Higher frequencies are thought to have a greater impact on elastoplastic deformation and surface roughness, resulting in a reduced friction force and thus a decrease in the required disassembly force.

It should be noted that the maximum achieved reduction in the disassembly force due to application of vibration was about 5%, which was significantly smaller than that reported in the literature, which was about 60% [3]. This difference could be attributed to variations in geometry, interference, and vibration parameters. For example, the diameter of the cylindrical pin used in the current study was 8 mm, while in the literature, it was less than 1 mm. Additionally, the contact area between the shaft and the hub was much larger in the current study, which increased the force exerted due to friction and decreased the effect of the applied oscillation waves.

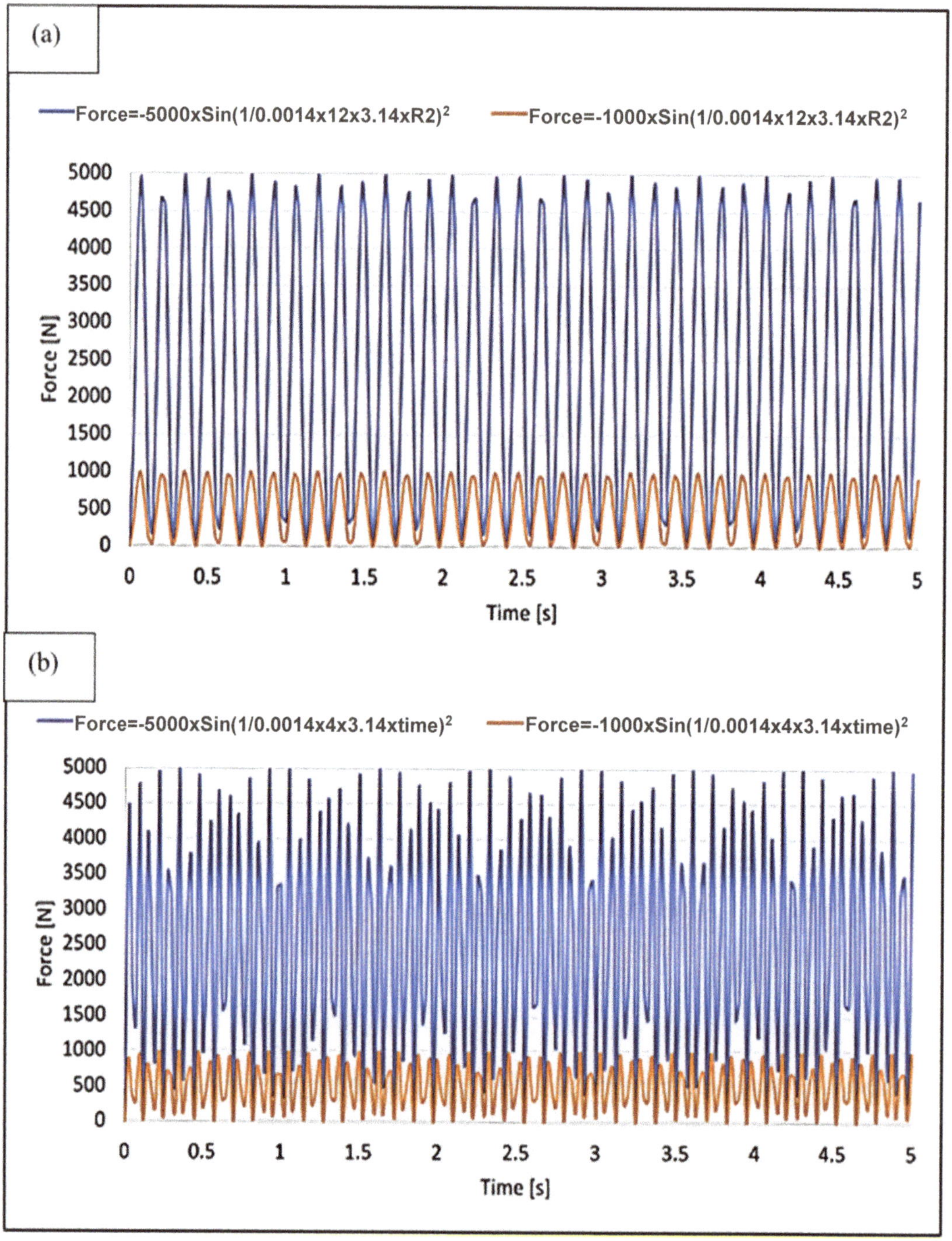

Figure 10. Oscillation force applied at (**a**) 700 Hz and (**b**) 1500 Hz.

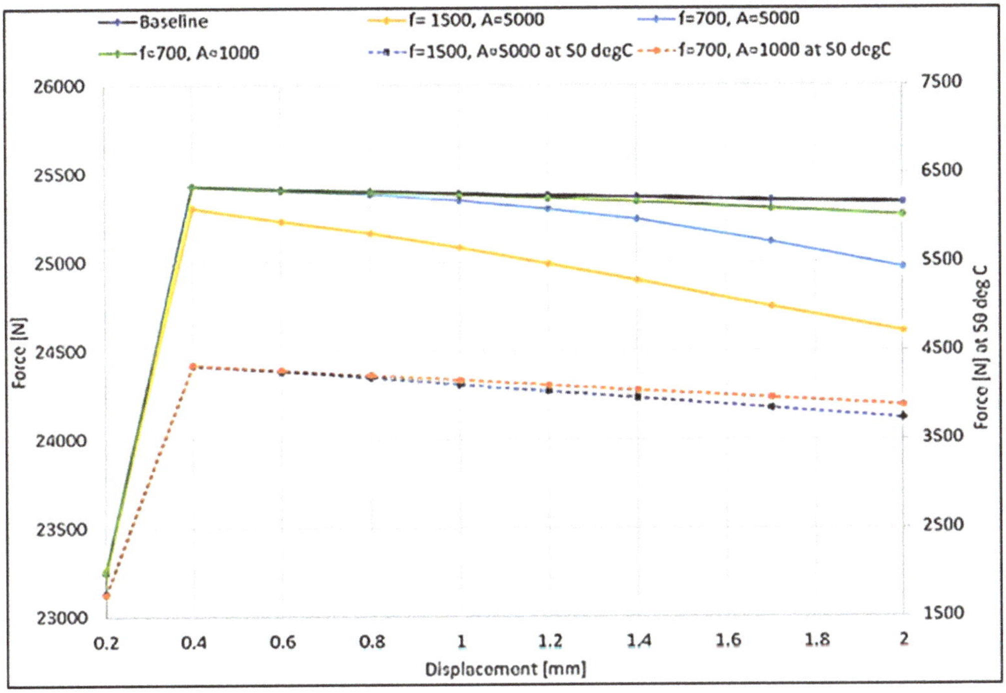

Figure 11. Disassembly force comparisons with low-frequency oscillations.

Further research could investigate the effect of higher frequency and an amplitude of oscillation waves, which would exert a higher impact of energy. Additionally, researchers could consider the thermal effect of oscillation waves, which could activate internal particles of the mating parts, enhance their relative motion, increase the temperature of the joint, and result in thermal softening, thereby decreasing the entire dynamic deformation resistance [24]. With the reduction in contact stress, the reduction in the disassembly force would be more drastic.

Regarding plastic strain, the maximum plastic strain was decreased from 1.57×10^{-2} to 1.83×10^{-3}, 1.72×10^{-4}, and 1.56×10^{-4}, respectively, when vibration waves of (frequency 700 Hz and amplitude 1000 N), (frequency 700 Hz and amplitude 5000 N), and (frequency 1500 Hz and amplitude 5000 N) were applied, which was equivalent to about an 88%, an 89%, and a 99% reduction in the maximum plastic strain. The average plastic stain also exhibited comparable lessening due to the application of the vibration waves. This softening effect of oscillation waves on plastic strain was previously suggested in the literature [24]. Moreover, the study showed that increasing both the amplitude and/or the frequency of the vibrations applied had a positive effect on the softening effect, which was confirmed by the current results.

Table 4 shows the plastic and thermal strains after disassembly at different amplitudes and frequencies of the oscillation waves applied. The table shows that applying vibration only does not reduce the plastic strain. The current study noted that the strain hardening effect was not taken into account in the model, which could lead to an overestimation of the reduction in the plastic strain after applying the vibration waves. Additionally, the plastic deformation at the micro-surfaces requires further investigation. Nevertheless, the results indicated that ND could still be implemented effectively with the aid of oscillation waves, even though the force reduction due to vibration was not significant. The effect of different vibration parameters needs to be investigated further.

Table 4. Plastic and thermal strains after disassembly at different amplitudes and frequencies of the oscillation waves applied.

Condition		Max. Plastic Strain	Average Plastic Strain	Max. Thermal Strain	Avg. Thermal Strain
Vibration Frequency (Hz) and Amplitude (N)	Thermal Flux (W/mm^2)				
Baseline		1.57×10^{-2}	1.42×10^{-3}	-	-
f = 700, A = 1000	-	1.83×10^{-3}	8.98×10^{-6}	-	-
f = 700, A = 5000	-	1.72×10^{-3}	6.33×10^{-6}	-	-
f = 1500, A = 5000	-	1.56×10^{-4}	8.30×10^{-6}	-	-
f = 700, A = 1000	1	1.59×10^{-4}	3.01×10^{-7}	3.39×10^{-4}	2.33×10^{-4}
f = 1500, A = 5000	1	1.42×10^{-4}	2.57×10^{-7}	3.39×10^{-4}	2.33×10^{-4}

Figure 11 and Table 4 showed that applying a heat flux of 1 W/mm^2 simultaneously with the oscillation waves resulted in a higher reduction in both plastic strain and breaking up force. This was suggested to be due to the decreased contact pressure. For instance, when a 1 W/mm^2 heat flux was applied concurrently with an oscillation wave of an amplitude of 5000 N and a frequency of 1500 Hz, the average plastic deformation and disassembly force (at a 2 mm displacement) were reduced to 2.57×10^{-7} and 3700 N, respectively. This reduction was equivalent to about 99% and 85%, respectively, compared to the case when the vibration wave of the same amplitude and frequency was applied alone (without thermal heating). These findings indicate that thermal disassembly could be more efficient than low-frequency oscillations with a suitable temperature increase.

4. Conclusions

This study aimed to explore strategies for improving the disassembly of parts for remanufacturing, with a focus on reducing disassembly force and frictional damage. An FEM model of an interference fit shaft-hub joint made from S235 steel was developed and validated, and potential solutions were investigated for dealing with deformation and corrosion wear. Thermal disassembly and vibration waves were identified as promising strategies for addressing these challenges. The key findings of the study are:

1. Plastic strain was found around the edges of the mating parts after applying a decoupling moment to the FEM. An application of a 1 W/mm^2 heat flux caused an 18% reduction in the breaking force, alongside a 3% decrease in plastic strain.
2. At a 2 W/mm^2 heat flux, the maximum temperature reached 220 °C, the breaking force reduced by 50%, but the plastic strain experienced a drastic boost as a result of the thermal strain.
3. Increasing the COF of S235 steel to 0.6 caused the disassembly force to increase by more than 250%.
4. An application of oscillation waves to the joint did not significantly affect the disassembly force, especially at the early stages of application. The maximum reduction obtained was about 5% with a vibration wave of 5000 N amplitude and 1500 Hz frequency.
5. The maximum plastic strain decreased by 99% when a vibration force of 1500 Hz and 5000 N amplitude was applied to the joint, possibly due to the softening effect of oscillation waves.
6. Simultaneously applying a heat flux of 1 W/mm^2 and vibration waves of 1500 Hz and 5000 N amplitude further decreased the breaking force by 85%, indicating the greater effectiveness of thermal-aided disassembly compared to vibration-assisted disassembly.

These findings provide valuable insights and tools for remanufacturing designers to facilitate disassembly, decrease disconnecting force, and ultimately reduce costs and the time required for the process. The results suggest that a combination of thermal disassembly and vibration waves can effectively reduce disassembly force and plastic strain, making it a promising approach for the remanufacturing industry. Further research is needed to explore the feasibility and safety of these methods on a larger scale and with different materials.

Author Contributions: Conceptualization, K.E.; Methodology, H.L.A.-Y. and K.E.; Formal analysis, H.L.A.-Y., H.H. and K.E. Investigation, H.L.A.-Y.; Resources, S.A., M.A.E.-S., M.A. and N.A.A.; Writing—original draft, N.A.A.; Writing—review & editing, S.A., H.H., M.A.E.-S., M.A., N.A.A. and K.E.; Visualization, H.L.A.-Y. and K.E.; Supervision, K.E.; Project administration, S.A., M.A.E.-S., M.A., M.M.Z.A. and N.A.A.; Funding acquisition, S.A., M.A.E.-S., M.M.Z.A., M.A. and N.A.A. All authors have read and agreed to the published version of the manuscript.

Funding: This work is funded by the Deanship of Scientific Research at Imam Mohammad Ibn Saud Islamic University (IMSIU) through Research Partnership Program no RP-21-12-04.

Institutional Review Board Statement: Not applicable.

Informed Consent Statement: Not applicable.

Data Availability Statement: The data presented in this study are available on request from the corresponding author. The data are not publicly available due to the extremely large size.

Acknowledgments: The authors extend their appreciation to the Deanship of Scientific Research at Imam Mohammad Ibn Saud Islamic University (IMSIU) for funding and supporting this work through Research Partnership Program no RP-21-12-04.

Conflicts of Interest: The authors declare no conflict of interest.

References

1. Yicong, G.; Shanhe, L.; Hao, Z.; Jianrong, T. A data-driven method of selective disassembly planning at end-of-life under uncertainty. *J. Intell. Manuf.* **2022**, *34*, 565–585.
2. Dan, Z.; Sen, L.; Xin, G.; Haihong, H. Nondestructive Surface Threshold Definition for Remanufacturing Disassembly of Interference Fit. *Int. J. Precis. Eng. Manuf.* **2022**, *19*, 1735–1743.
3. Dieudonné, E.; Offole, F.A.; Nkongho, A.J.A.; Ngayihi, A.C.V.A.; Njomoue, P.A.A.; Zanga, C.N. A study on the experimental investigation of low frequency vibration wave assisted disassembly of press-fit joints. *J. Manuf. Process.* **2020**, *49*, 70–81. [CrossRef]
4. Hu, J.; Zhang, K.; Cheng, H.; Qi, Z. An experimental investigation on interfacial behavior and preload response of composite bolted interference-fit joints under assembly and thermal conditions. *Aerosp. Sci. Technol.* **2022**, *103*, 105917. [CrossRef]
5. Abuzied, H.; Senbel, H.; Awad, M.; Abbas, A. A review of advances in design for disassembly with active disassembly applications. *Eng. Sci. Technol. Int. J.* **2021**, *23*, 618–624. [CrossRef]
6. Zhu, J.; He, M. Product sustainable design information model for remanufacturing: A case study in the construction equipment industry. *Int. J. Adv. Manuf. Technol.* **2022**. [CrossRef]
7. Taghizadeh, H.; Chakherlou, T.N. Fatigue behavior of interference fitted Al-alloy 7075-T651 specimens subjected to bolt tightening. *Proc. Inst. Mech. Eng. Part J. Mater. Des. Appl.* **2019**, *233*, 1879–1893. [CrossRef]
8. Radi, E.; Lanzoni, L.; Strozzi, A.; Bertocchi, E.J.A.M.M. Shaft-hub press fit subjected to bending couples: Analytical evaluation of the shaft-hub detachment couple. *Appl. Math. Model.* **2017**, *50*, 135–160. [CrossRef]
9. Hammond, R.; Amezquita, T.; Bras, B. Issues in the automotive parts remanufacturing industry: A discussion of results from surveys performed among remanufacturers. *Int. J. Eng. Des. Autom.* **1998**, *4*, 27–46.
10. Liu, Y.; Mol, J.; Janssen, G.J.S.M. Corrosion reduces wet abrasive wear of structural steel. *Scr. Mater.* **2015**, *107*, 92–95. [CrossRef]
11. Soh, S.; Ong, S.; Nee, A.J.A.A. Design for assembly and disassembly for remanufacturing. *Assem. Autom.* **2016**, *36*, 12–24. [CrossRef]
12. Mok, H.; Kim, H.; Moon, K.J.C. Disassemblability of mechanical parts in automobile for recycling. *Comput. Ind. Eng.* **1997**, *33*, 621–624. [CrossRef]
13. Pedersen, N.L. Optimization of contact stress distribution in interference fit. In Proceedings of the 11th World Congress of Structural and Multidisciplinary Optimization, Sydney, Australia, 7–12 June 2015.
14. Zhou, D.; Xu, Y.; Gao, X.; Huang, H.-H. Experimental study on the reduction effect of pit texture on disassembly damage for interference fit. *Res. Sq.* **2020**, *49*, 70–81. [CrossRef]
15. Wang, F.; Zhang, S.; Hao, S.; Shu, H. The Nondestructive Disassembly Method of Interference Fit of Sleeve-Base Structure in the Case of Cooling Excitation. In Proceedings of the 6th International Conference on Sustainable Design and Manufacturing (KES-SDM 19), Budapest, Hungary, 4–5 June 2019.
16. Shaheen, M.A.; Foster, A.S.; Cunningham, L.S.; Afshan, S. Behaviour of stainless and high strength steel bolt assemblies at elevated temperatures—A review. *Fire Saf. J.* **2020**, *113*, 102975. [CrossRef]

17. Sun, J.; Nitschke-Pagel, T.; Dilger, K. Influence of strain-hardening models and slopes on the predicted residual stresses in structural steel S235 weldments. *J. Mater. Res. Technol.* **2022**, *19*, 4044–4062. [CrossRef]
18. Bengeri, M.; Mack, W. The influence of the temperature dependence of the yield stress on the stress distribution in a thermally assembled elastic-plastic shrink fit. *Acta Mech.* **1994**, *103*, 243–257. [CrossRef]
19. Essola, D.; Amba, J.C.; Ngayihi Abbe, C.V.; Nkongho Anyi, J.I. Enhancement of metallic machine parts mechanical properties by the use of vibratory processing for oxide coated films formation and MoS2 solid lubricant coating deposit. *Int. J. Mech. Mater. Eng.* **2019**, *14*, 8. [CrossRef]
20. Fridman, H.D.; Levesque, P. Reduction of static friction by sonic vibrations. *J. Appl. Phys.* **1959**, *30*, 1572–1575. [CrossRef]
21. Gutowski, P.; Leus, M. The effect of longitudinal tangential vibrations on friction and driving forces in sliding motion. *Tribol. Int.* **2012**, *55*, 108–118. [CrossRef]
22. Mikolainis, J.; Bakšys, B. Experimental investigation of interference fit connection of mechanical components. *J. Vibroeng.* **2012**, *14*, 73–78.
23. Langenecker, B. Effects of ultrasound on deformation characteristics of metals. *Ultrasonics* **1966**, *13*, 1–8. [CrossRef]
24. Wang, C.; Liu, Y.; Guo, B.; Shan, D.; Zhang, B.J.M. Acoustic softening and stress superposition in ultrasonic vibration assisted uniaxial tension of copper foil: Experiments and modeling. *J. Mater. Des.* **2016**, *112*, 246–253. [CrossRef]
25. Madej, J.; Śliwka, M. Analysis of interference-fit joints. *Mater. Today Proc.* **2021**, *11*, 11428. [CrossRef]
26. Croccolo, D.; De Agostinis, M.; Vincenzi, N. Design of hybrid steel-composite interference fitted and adhesively bonded connections. *J. Adhes.* **2012**, *88*, 19–25. [CrossRef]
27. *ISO 286-1:1988;* Specification for ISO System of Limits and Fits, Part 1, Bases of Tolerances, Deviations and Fits. ISO: Geneva, Switzerland, 1993.
28. İrsel, G. Study of the microstructure and mechanical property relationships of shielded metal arc and TIG welded S235JR steel joints. *Mater. Sci. Eng.* **2022**, *830*, 142320. [CrossRef]
29. Sen, S.; Aksakal, B. Stress analysis of interference fitted shaft–hub system under transient heat transfer conditions. *J. Strain Anal. Eng. Des.* **2004**, *39*, 407–417. [CrossRef]
30. En 10025-2 Grade S235JR As-Rolled Condition (+ar)- Low Carbon Steel. Available online: https://matmatch.com/materials/minfm33118-en-10025-2-grade-s235jr-as-rolled-condition-ar- (accessed on 11 August 2022).
31. Gassama, D.; Diagne, A.A.; Yade, I.; Fall, M.; Faty, S. Investigations on the corrosion of constructional steels in different aqueous and simulated atmospheric environments. *Bull. Chem. Soc. Ethiop.* **2015**, *1*, 99–109. [CrossRef]
32. Kocanda, D.; Jurczak, W.; Lunarska, E.; Swiatek, K. Effect of frictional-mechanical treatment of the low alloy steels on some exploitation properties in sea water simulated solution. *Solid State Phenom. Trans. Tech. Publ.* **2016**, *250*, 56–60. [CrossRef]
33. Chu, N.; Ning, Y.; Yu, L.; Huang, Q.; Wu, D. A high-resolution and low-frequency acoustic beamforming based on Bayesian inference and non-synchronous measurements. *IEEE Access* **2020**, *8*, 82500–82513. [CrossRef]
34. Antoni, N. Contact separation and failure analysis of a rotating thermo-elastoplastic shrink-fit assembly. *Appl. Math. Model.* **2013**, *37*, 2352–2363. [CrossRef]

machines

MDPI

Article

Parametric Modeling of Curvic Couplings and Analysis of the Effect of Coupling Geometry on Contact Stresses in High-Speed Rotation Applications

Chara Efstathiou *, Ioanna Tsormpatzoglou and Nikolaos Tapoglou

Department of Industrial Engineering & Management, International Hellenic University, Sindos Campus, 57400 Thessaloniki, Greece

* Correspondence: efhara@gmail.com

Abstract: Curvic couplings are used in applications demanding high positional accuracy and high torque transmission; therefore, improving their design and enhancing their load-carrying capacity is crucial. This study introduced the kinematic model *Curvic3D*, which was developed to produce the accurate geometry of both members of a curvic coupling using a CAD system. The model enabled the complete parametrization and customization of the coupling design using important geometric parameters. The couplings produced using *Curvic3D* were then imported into a finite element analysis model also developed as part of this study. A detailed analysis of the stresses developed on the teeth of the concave and convex parts provided information about the behavior of the coupling under different loading conditions. Finally, a series of geometric parameters, such as the number of teeth, the number of half pitches, the root fillet radius, and gable angle were examined as to their influence on the load-carrying capacity of the curvic coupling. The study concluded that all the examined parameters have a significant effect on the tooth flank and root area stresses.

Keywords: curvic coupling; CAD; manufacturing modeling; simulation; FEA

1. Introduction

1.1. Curvic Coupling Applications

Curvic couplings are important components used in many industrial applications and especially in joining two shafts or two sections of a shaft. They are well-suited to applications demanding high precision, reliability, and high torque transmission capabilities. Curvic couplings are commonly used in the following sectors: 1. The aerospace industry, in jet engines, gas turbine power plants, and helicopter rotor systems. 2. Power generation, in gas and steam turbines, enabling the transmission of torque from the turbine to the generator. 3. Marine propulsion, in ship propulsion shafts and propellers. 4. Heavy machinery, such as mining equipment, industrial compressors and large-scale machine tools. 5. The defense industry, including military aircraft, armored vehicles and naval vessels. 6. The oil and gas industry, in oil and gas drilling and production machinery. They enable torque transmission in drilling rigs, pumps and compressors. 7. Robotics and automation, where precise motion control is necessary. They also find applications in robotic arms, CNC machines, and other automated equipment.

1.2. Curvic Coupling Geometry and Manufacturing

Curvic couplings are mechanical connections utilized for the connection of two rotating components and transmitting torque between them. They are designed to provide rigid connection, ensuring the accurate alignment of the shafts and high torque transmission. Additionally, a curvic coupling is advantageous in that its positional accuracy in both axes actually improves over time as opposed to degrading [1]. A curvic coupling is ring-shaped and consists of two members: the convex and the concave part. In most cases, the two members are clamped together with bolted connections. Both components have curved teeth placed circumferentially on the face of the part. This curvature is formed as a result of the machining process kinematics and tool geometry. Curvic couplings are most commonly machined with face-mill cutters or cup-type grinding wheels [2]. One member of the coupling is machined with the outer edge of the cutter, resulting in a concave tooth form, whereas the other member is machined with the inner edge of the cutter, forming a convex tooth, as shown in Figure 1. The kinematics of the process is relatively simple and is realized in the following steps. Firstly, the cutter engages with the left and right tooth flank feeding into the coupling until the final depth of the coupling slot is reached. The cutter is then retracted from the workpiece and remains at a certain distance until the workpiece rotates one pitch around its axis. Finally, the first step is repeated and the cutter is fed into the work gear at the next indexing position to machine the next slot of the coupling. This manufacturing process is a single-indexing process; thus, the slots are cut two at a time, and the manufacturing of a complete part of the curvic coupling is completed when all slots of the part are formed.

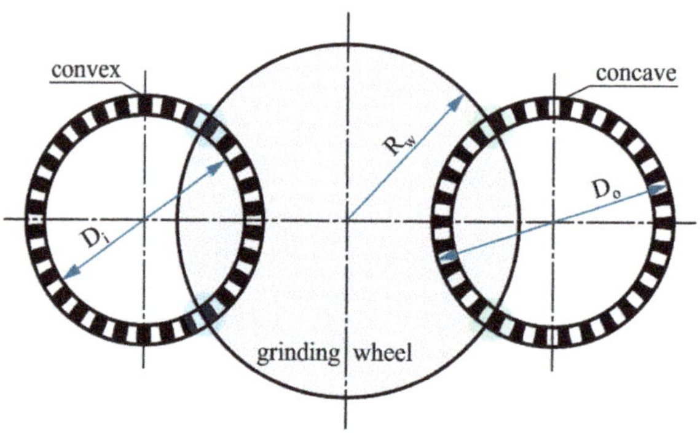

Figure 1. Curvic coupling manufacturing principle.

Figure 2 presents the basic geometry of the curvic coupling cutting tooth profile. The main geometric parameters affecting the geometry of the coupling teeth include the tooth addendum h_a, tooth dedendum h_d, normal pressure angle a_n, chamfer height h_c, chamfer angle a_c, gable angle a_g, and tooth root radius r_t.

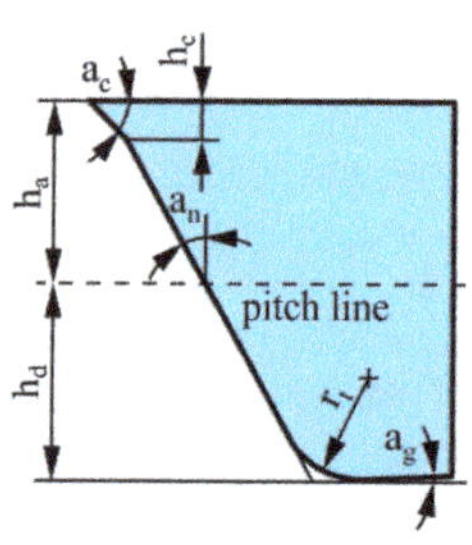

Figure 2. Curvic coupling cutting tooth profile.

2. State of the Art and Contribution of the Present Study

Research on curvic couplings and specifically curvic coupling manufacturing is relatively limited to the industrial sector. In their article, Gleason Works [2] provided fundamental knowledge on curvic couplings' design. The basic types of curvic couplings were listed, and brief descriptions of the respective manufacturing processes were provided. The authors also made useful suggestions for the geometry of fixed curvic couplings. The most important contribution of their work was the description of the complete design procedure to determine the necessary geometry of a curvic coupling to transmit a certain load. Richardson et al. [3] developed a three-dimensional and a two-dimensional finite element model for the simulation of the contact behavior of curvic couplings. The study aimed to compare the two simulation approaches and determine whether two-dimensional modeling is sufficient to analyze the contact of curvic couplings. A study to validate the three-dimensional finite element contact developed by Richardson et al. was presented in [4]. The finite element method results were evaluated in comparison with the results obtained from a photoelastic test. Rensis et al. [5] developed a three-dimensional axisymmetric curvic contact model to predict the maximum stresses in three loading cases. In his work on gas turbine engines, Boyce [1] described the role of couplings in power transmission between two shafts of an engine. Among others, the author discussed the use of curvic couplings in such setups. A three-dimensional finite element model was developed by Jiang et al. [6] to analyze the contact stresses of the curvic coupling in a gas turbine under a blade-off load condition. A methodology for the design of Hirth ring couplings which are used in the machine tool industry was provided by Croccolo et al. [7]. First, the standard formulas currently used in the industry were listed. Then, the forces generated in Hirth couplings were calculated with a new analytical method taking into account the role of friction. The proposed equations were also experimentally validated on a machine tool rotary table. Zhang et al. [8] presented a modified analytical method to calculate the equivalent stress on the double-row curvic coupling teeth, taking the deep beam bending effect into account. A finite element model was also presented to analyze the contact stress of the curvic couplings under different loads. The study concluded that the bolt preload has the greatest impact on the contact stress, while the rotating speed of the shaft reduces the contact stress. A new type of large curvic coupling gear consisting of a large gear and a curvic coupling was presented by Jung et al. [9]. Finite element analysis simulations were conducted to determine the maximum Von Mises stress developed on the model under two different external loads. Nielson [10] studied the potential of using CMM as a means of investigating the contact pattern of curvic couplings. In the field of the simulation of manufacturing processes, a series of studies have shown the potential of using CAD-based simulation models in complex manufacturing processes [11,12]. Li et al. [13] proposed a curvic coupling design with a double circular arc root fillet to improve the stress concentration on the tooth root. Huang et al. [14] proposed a method to calculate the machine setting parameters for curvic coupling manufacturing on a spiral bevel gear grinding machine. Pisani et al. [15] investigated the behavior of curvic couplings using two- and three-dimensional boundary finite element models. The two approaches were compared as to the required effort for the model generation, the results' accuracy levels, and the post-processing capability. A method for the calculation of the contact and bending stiffness of a curvic coupling was developed by Yu et al. [16]. Kim et al. [17] presented a novel approach in curvic coupling modeling combining a Greenwood–Williamson contact model with a 3D solid element model. A study was also performed to investigate the effect of several parameters on vibration and stress. Yang et al. [18] investigated the stiffness of curvic couplings in order to determine the tensile–compressive stiffness of the coupling. In their investigation, they considered the stiffness characteristics of the coupling with uniform and non-uniform load distribution. A novel mechanical model considering the curvic coupling stiffness weakening in various loading conditions, such as shearing, compression, bending, and torsion, was developed by Liu et al. [19].

The present study introduces the first model for the kinematic simulation of curvic couplings' manufacturing that accurately produces the convex and concave tooth geometry. Moreover, finite element analysis is utilized for the investigation of the effects of several geometric features of the two members on the load-carrying capacity of the coupling. A complete platform for the design, analysis of in-use performance, and manufacturing has been developed, allowing the end user to optimize the design and performance of curvic couplings.

3. CAD Model

Curvic3D is a CAD-based simulation model developed as part of this study. The algorithm simulates the process kinematics so the curvic tooth flank solid geometries are produced as output. This simulation approach aims to enable the accurate parametric modeling of curvic couplings. Figure 3 shows a flowchart of the simulation process. Modeling and simulation procedures are implemented in the following steps:

I: Calculation and modeling of the blank curvic coupling geometries.

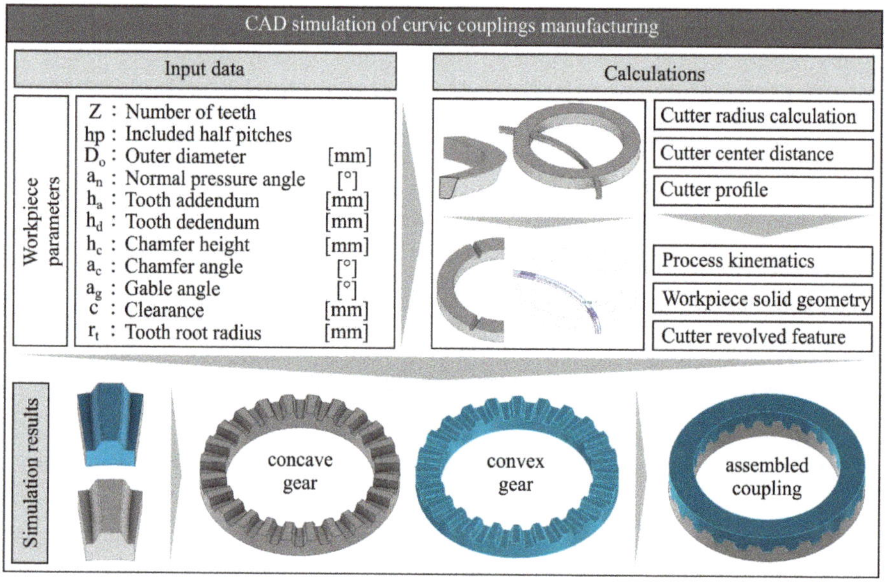

Figure 3. CAD-based simulation model flowchart.

The calculation process begins with the modeling of the two blank geometries for the convex and the concave component of the coupling. The blank geometries consist of two simple rings, the inner diameter of which is the inner diameter of the curvic coupling, while the outer diameter is the outer diameter of the curvic coupling. The area between the inner and outer diameter forms the face width of the curvic coupling tooth. As stated in [2], the inner diameter of the coupling should be equal to or greater than 75% of the outer diameter.

II: Tool profile and cutter geometry calculation.

Afterwards, the tooth profile geometry is calculated and drawn according to DIN 3972 standard [20]. The cutter radius and cutter center distance are calculated and considered in the modeling process.

III: Simulation of the process kinematics. Tool trajectory creation.

In order to obtain the solid geometry of the cutting tool, the process kinematics must be integrated with the tooth profile and cutter geometry. The kinematics of the process consists of a rotation of the cutter around its axis and feed of the cutter towards the workpiece. Parameters such as the cutter radius and cutter center distance are taken into consideration, and the solid geometry of the cutting tool is formed, as shown in the upper right part of Figure 3, as a result of the combination of process kinematics, tooth profile, and cutter geometry.

IV: Calculation of the convex and concave components' geometry.

The final step of the simulation includes the interaction of the cutting tool and the two workpieces. Two subassemblies are formed; the first one consists of the cutting tool and the concave component, and the second one includes the cutting tool and the convex component. Using Boolean operations, such as Boolean subtraction, the simulated curvic tooth surface can be obtained. The algorithm is configured so that the final finishing pass of the machining operation is implemented; therefore, the cutting tool is placed directly at the final depth of the curvic slot. Nonetheless, the algorithm supports rough machining operation with the tool performing multiple passes until the final slot depth. This way, the undeformed chip geometries can also be obtained, enabling the calculation and analysis of the cutting forces.

4. Finite Element Analysis and Simulation Model

Aiming to achieve the analysis of the contact behavior and developed stresses of curvic couplings with variable geometries under various loading conditions, a finite element simulation model was developed. The solid models of the couplings were obtained using the CAD model *Curvic3D*. Three-dimensional modeling was preferred over two-dimensional as the complex geometry of a curvic coupling can be represented more accurately. The model aims to simulate the contact conditions and contact stresses for a curvic coupling in high-rotating speed applications such as high-pressure turbine shafts. The primary loads that couplings withstand in such applications are considered in the model, and all necessary assumptions are made, as described in the following paragraphs.

4.1. Solid Models and Materials

Figure 4 shows two indicative solid models of the two members of a curvic coupling used in the simulation. The assembled coupling is obtained from *Curvic3D*, and it can be used directly as is. A specific material has to be assigned to both members of the coupling. In the present investigation, AISI 4340 annealed steel was assigned to the curvic coupling throughout the study.

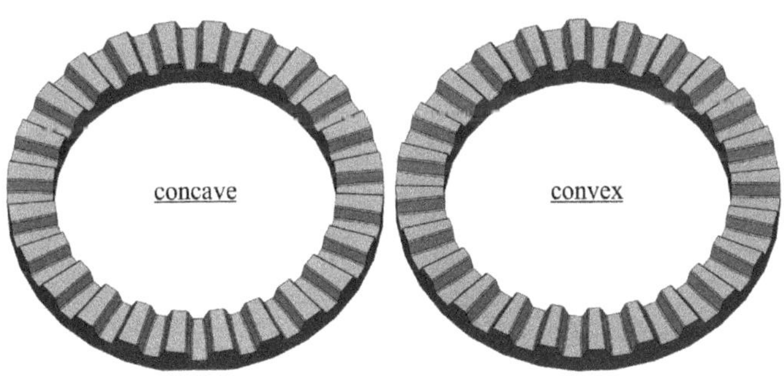

$Z = 24$, $hp = 17$, $D_o = 100$ mm, $w = 12.5$ mm, $\alpha_g = 0°$, $a_n = 20°$,
$h_c = 0.391$ mm, $h_a = 1.696$ mm, $h_d = 2.13$ mm

Figure 4. Curvic coupling solid models.

AISI 4340 is a nickel–chromium–molybdenum alloy steel, known for its toughness and strength, and it is used, among other applications, in gear manufacturing. Table 1 shows the specific material properties.

Table 1. Material properties of AISI 4340 annealed steel.

Property	Value	Units
Density	7850	kg/m^3
Elastic modulus	205	GPa
Yield strength	470	MPa
Specific heat	475	J/Kg·K

4.2. Global Mesh and Local Mesh Controls

The selected element type for both global and local mesh is the ten-node tetrahedral solid element. Two regions of mesh were assigned to both parts. The local mesh control with a 0.4 mm element size was applied to the coupling tooth surface to provide more accurate results since this was the region of interest. The global mesh applied to the rest of the solid bodies of both members in order to control the computational time had a 1.5 mm element size. Figure 5 shows the two types of mesh applied to each member.

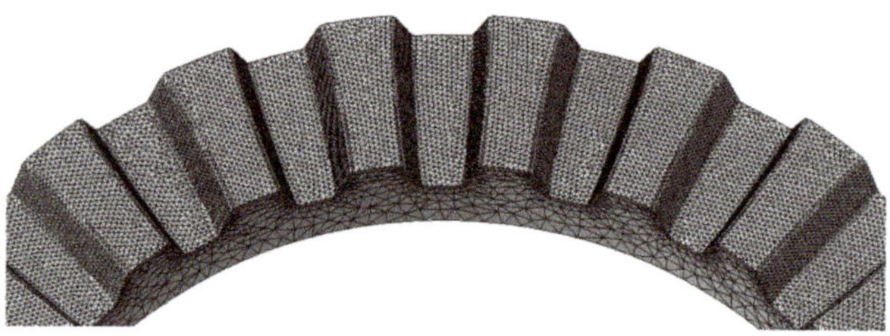

Figure 5. Global mesh and tooth surface mesh control on the concave part.

4.3. Boundary Conditions

Figure 6 presents the final boundary conditions of the model. The displacements and loads were defined based on the literature [8,9] as well as a series of simulations.

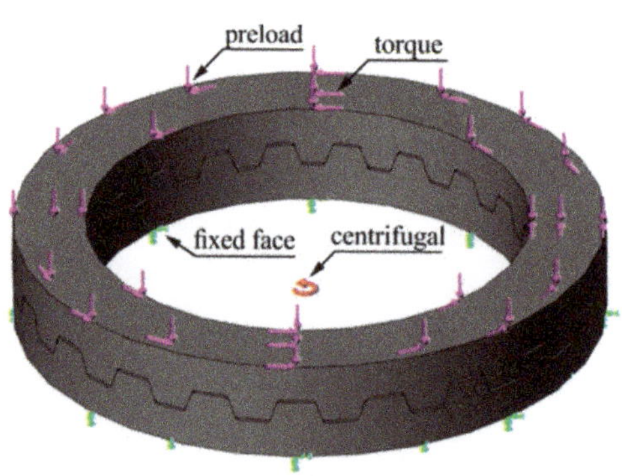

Figure 6. Finite element model boundary conditions.

4.3.1. Loads

Centrifugal force: The centrifugal force applied on the curvic coupling results from the rotating speed of the shaft, which equals 10,000 rpm.

Torque: Equation (1) gives the separating force F_s acting on the curvic coupling as a result of the torque. Equation (2) provides the minimum clamping force F_c to counterbalance the action of separating force F_s. Equations (3) and (4) provide the stresses σ_t and σ_c developed due to the act of clamping force and torque. The allowable torque is calculated according to Equation (5), which gives the equivalent stress on the coupling teeth caused when the coupling is subjected to the bolt preload (clamping force) F_c and the torque T.

$$F_s = \frac{T}{A} \times \tan a_n \quad (1)$$

$$F_c = 1.5 \times \frac{T}{A} \times \tan a_n \quad (2)$$

$$\sigma_t = \frac{T}{A \times Z \times w \times h_0} \quad (3)$$

$$\sigma_c = \frac{F_c}{2 \times Z \times w \times h_0 \times \tan a_n} \quad (4)$$

$$\sigma_{eq} = \frac{F_c}{2 \times Z \times w \times h_0 \times \tan a_n} + \frac{T}{A \times Z \times w \times h_0} \quad (5)$$

According to Equation (5), setting the maximum allowable stress on the teeth at $\sigma_a = \sigma_y/s_f = 156.7$ MPa with the safety factor at $s_f = 3$, the maximum allowable torque value is calculated to be T = 4642 N·m. This torque is applied to the model for all the simulations performed in this study.

Clamping force: To define the displacement boundary conditions of the convex gear, a series of simulations was conducted to determine the necessary preload force to compensate for all the separating forces acting on the curvic coupling. The minimum preload clamping force to prevent the separation of the two members should exceed the sum of all the separating forces acting on the coupling. According to Gleason Works [2], the clamping force should be at least 1.5 2 times the sum of all separating forces acting on the curvic coupling teeth. In this study, several simulations were performed to determine the necessary clamping force so the curvic coupling was not disassembled. For a given centrifugal force resulting from the rotational speed N = 10,000 rpm and torque T = 4642 N·m, the minimum clamping force was defined as $F_c = 42$ kN. Applying this clamping force to the coupling, the distance between the two members during simulation was not increased by more than 15 μm.

4.3.2. Displacements

As shown in Figure 6, the axial and angular displacements at the back face of the concave member are constrained to 0 (fixed face), a necessary assumption made to avoid the excessive displacement of the model and prevent the simulation from failing. In operating conditions, the displacements' constraints would be shared between the convex and the concave part. The displacements of the convex member are only controlled via the clamping force and are not subject to any other constraints. Furthermore, an investigation to determine the suitable displacement boundary conditions was carried out, examining the behavior of the model under different displacements' constraints. The above described displacements were selected because they better represent the actual curvic coupling operation and have also been used by other researchers in the literature [8,9].

4.4. Interaction Conditions

The no-penetration contact condition was applied globally to the model. Under this contact condition, the two parts/models of the curvic assembly behave as two separate solid bodies that interact but cannot penetrate each other. The friction coefficient between the two parts was set to $\mu = 0.15$.

4.5. Results

An analysis of the coupling behavior under different loading conditions was carried out prior to determining the final loading state that would be applied on the curvic coupling. Two loading cases were examined: 1. assembly, where only the clamping force is applied to the model, and 2. operation, where preload force, centrifugal force, and torque are loading the coupling as in normal operation. The results showed that loading the curvic coupling with only the clamping force was the most demanding case with respect to the developed stresses as the maximum Von Mises stress reached $\sigma_{v,max}$ = 469 MPa, which is practically the yield strength of the material. In the second loading case of operation where the coupling was subjected to all three loads, the maximum Von Mises stress did not exceed $\sigma_{v,max}$ = 357 MPa.

A baseline finite element simulation was performed using the boundary conditions presented above and was used as a reference for the subsequent parameters investigation. The solid model of the curvic coupling was obtained from *Curvic3D*, and the geometric features of the coupling are shown in Figure 7. As can be seen in the figure, the stresses are distributed evenly on the loaded tooth surface of both concave and convex members. The maximum Von Mises stress occurred along the tooth root of the convex gear. While the opposite tooth surface was not loaded, the opposite root area carried some load but the stress remained at low levels.

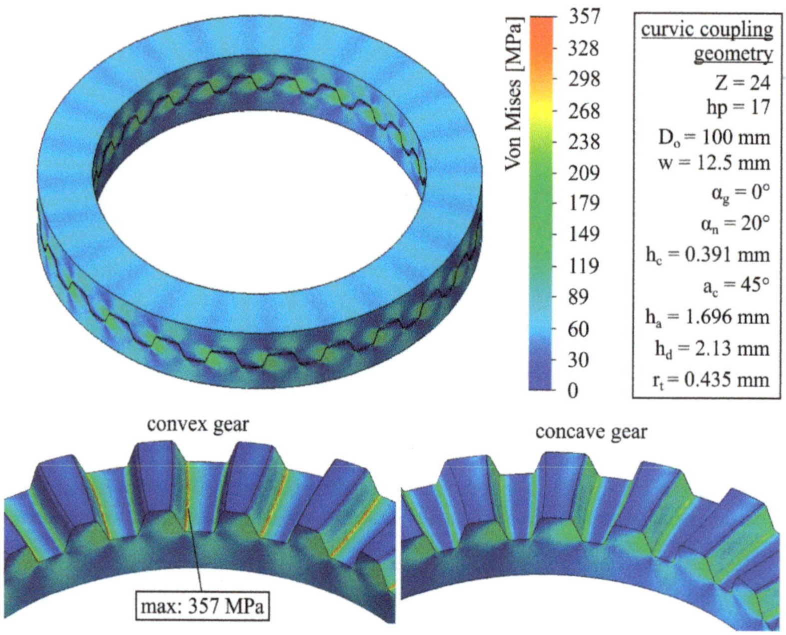

Figure 7. Von Mises stress plot.

Since the maximum Von Mises stress of $\sigma_{v,max}$ = 357 MPa is lower than the yield strength of the material (σ_y = 470 MPa) and the minimum factor of safety is FOS = 1.316, the results were reasonable and could be used as a reference for the subsequent investigation. Figure 8 presents the factor of safety plot for the convex member of the coupling. As can be observed, the stresses developed for the most part of the curvic correspond to an FOS = 3, except for a narrow region at the root of the loaded flank of the tooth where the safety factor drops at lower values. These results verify the calculation procedure using Equations (1)–(5), considering that these equations are only used as an approximation of the actual loading conditions; hence, many assumptions and simplifications are made.

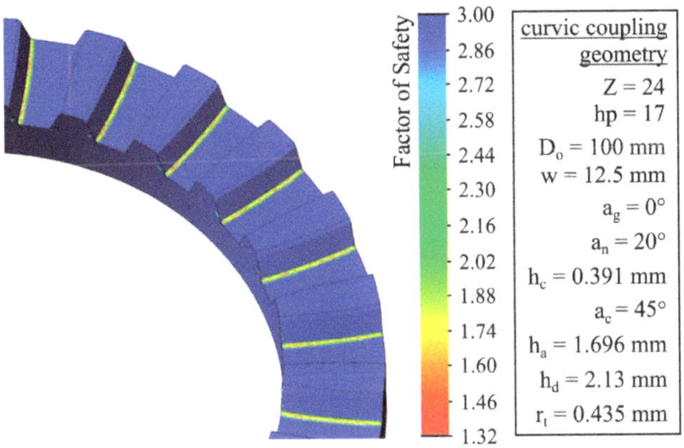

Figure 8. Factor of safety plot.

Figure 9 shows the displacement plot of the curvic coupling. As mentioned above, the displacements of the back face of the concave part are constrained to 0, and this is also noticeable in the figure. The maximum displacements are observed on the convex member and stay below 12 μm due to the adjustment at the clamping load, which was determined in order to counterbalance the separating forces acting on the coupling.

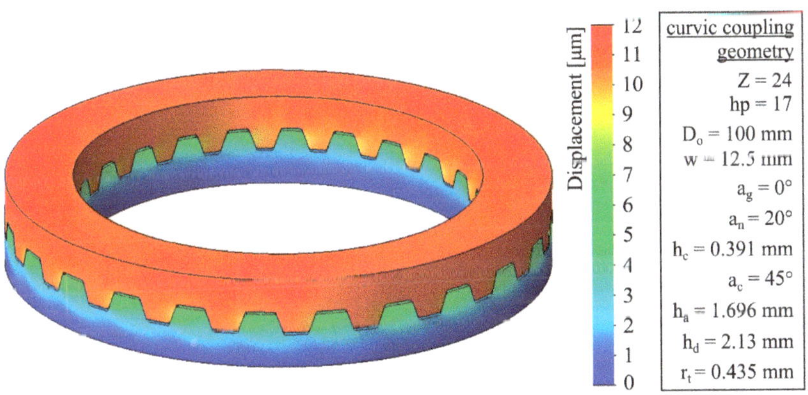

Figure 9. Displacement plot.

5. Investigation of the Effect of Geometric Parameters on the Contact Stresses

Following the development of the FEA simulation model and the initial assessment of the simulation results, a series of simulations was carried out to investigate the effect of several coupling geometric parameters on the developed Von Mises stresses. For each simulation case, a separate curvic coupling was modeled using *Curvic3D* and was subsequently embedded in the finite element model.

5.1. Effect of the Number of Teeth

For the examination of the effect of the number of teeth, three simulation cases with varying numbers of teeth—Z = 21, 24 and 27—showed that the maximum developed Von Mises stress decreases from $\sigma_{v,max}$ = 394 MPa to $\sigma_{v,max}$ = 330 MPa as the number of teeth increases. Figure 10 shows the values of maximum Von Mises stress obtained from the three simulation cases. All of the maximum stress values were developed at the tooth root of the convex gear. The dramatic effect of the number of teeth on the Von Mises stress is explained by the fact that as the number of teeth increases, the load is distributed amongst more teeth, leading to a decrease in overall stresses.

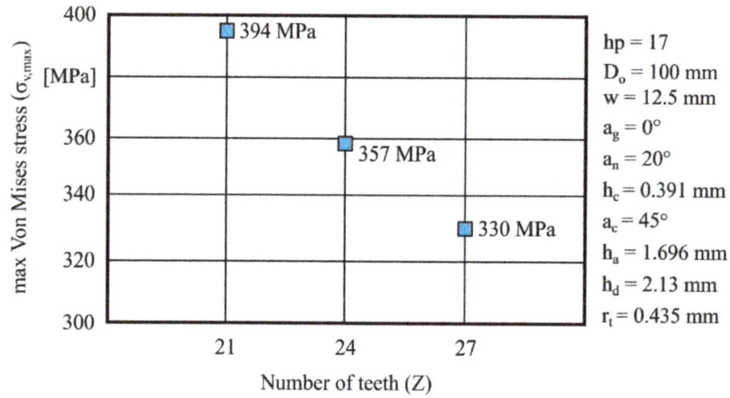

Figure 10. Effect of the number of teeth (Z) on the developed Von Mises stress.

5.2. *Effect of the Number of Half Pitches*

The number of half pitches included between the two tooth flanks that the cutter machines is also an important parameter affecting the geometry and therefore the strength of the curvic couplings. As shown in Figure 11, the increase in half pitches decreases the maximum developed Von Mises stress from $\sigma_{v,max}$ = 363 MPa to $\sigma_{v,max}$ = 319 MPa. This can be explained due to the fact that the maximum stress in all simulation cases develops at the convex gear tooth root, the tooth thickness of which increases with the increase in half pitches.

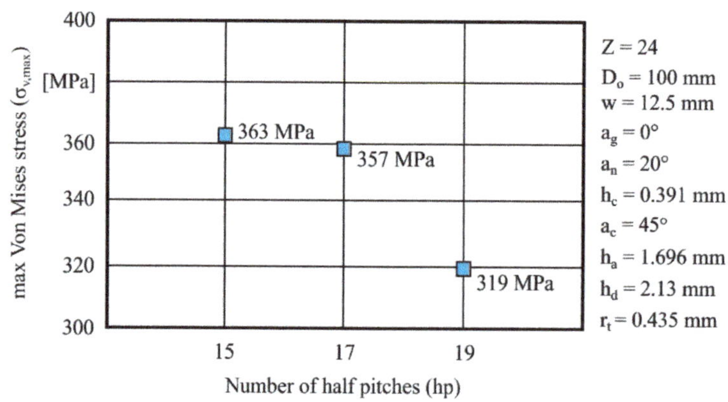

Figure 11. Effect of the number of half pitches (hp) on the developed Von Mises stress.

5.3. *Effect of Tooth Root Radius*

Analyzing the effect of tooth root radius on the development of Von Mises stresses, three cases were examined in which the tooth radius was r_t = 0.235, 0.435, and 0.635 mm. In all three cases, the maximum stress occurred on the root of the convex member; therefore, the change in root radius had a great impact on the maximum Von Mises stress which dropped from $\sigma_{v,max}$ = 388 MPa to $\sigma_{v,max}$ = 303 MPa, as shown in Figure 12. The increase in the root radius enhanced the strength of the coupling; thereby, the maximum stress decreased dramatically.

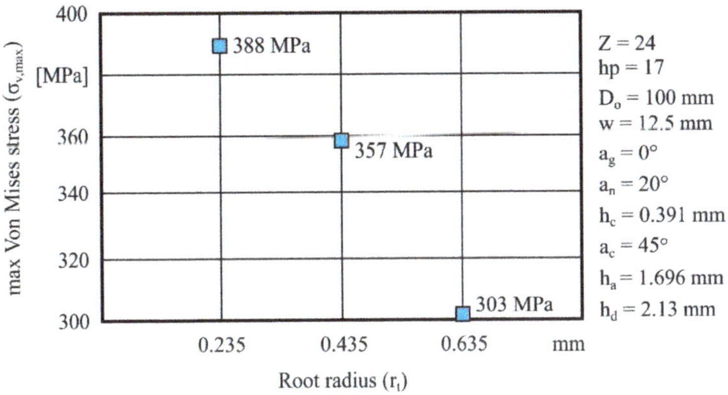

Figure 12. Effect of the tooth root radius (r_t) on the developed Von Mises stress.

5.4. *Effect of the Gable Angle*

A useful parameter that many curvic coupling manufacturers take into consideration is the gable angle. A total of seven simulations were executed to determine the effect of gable angle on the development of Von Mises stress on the coupling. The results revealed an interesting relation between the increase in gable angle and maximum Von Mises equivalent stress. As can be seen in Figure 13, the maximum Von Mises stress remains constant at about 360 MPa from $a_g = 0°$ until the gable angle reaches 1.5° and then slightly increases to reach $\sigma_{v,max}$ = 376 MPa at $a_g = 3.5°$. At $a_g = 4°$, the maximum Von Mises stress value drops abruptly close to $\sigma_{v,max}$ = 305 MPa and remains constant for the rest of the simulation cases. Hence, it can be concluded that, for this particular curvic coupling geometry, values of the gable angle between $a_g = 4°$ and 5° are optimum as they would result in reduced stresses. Values of the gable angle above 5° were not examined since the produced geometry would not lead to an allowable assembly.

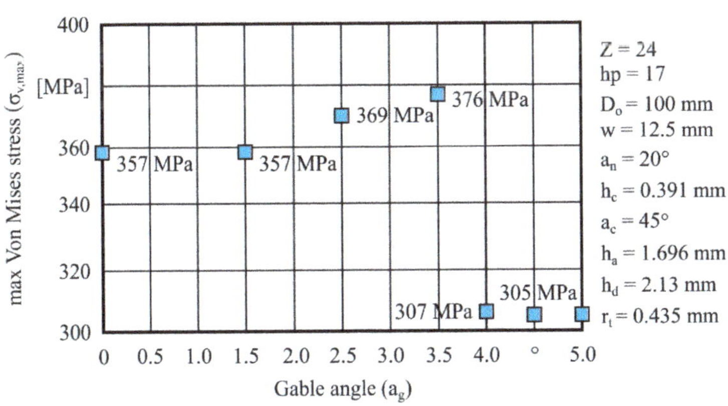

Figure 13. Effect of the gable angle (a_g) on the developed Von Mises stress.

6. Conclusions

The CAD model *Curvic3D* for the parametric three-dimensional modeling of curvic couplings was developed as part of this study. The model simulates the manufacturing process kinematics achieving the automatic modeling of the solid geometries of both convex and concave members of the coupling, using custom-defined parameters. A finite element analysis was also carried out to study the stresses developed on the tooth contact surface during operation at high rotating speed. Several case studies were simulated to investigate the influence of different geometric parameters on the developed Von Mises stress. The solid geometries of the curvic couplings were obtained from *Curvic3D* and examined with FEA for their load-carrying capacity. The number of teeth, the number of half pitches, the tooth root radius, and the gable angle were analyzed as to their effect on the load-carrying capacity of the curvic coupling. All of these parameters were found to have a great impact on the strength of the coupling, and the results were discussed in detail. The tooth root radius had the strongest impact on the developed stresses, followed by the number of teeth and the gable angle. All of the parameters showed a standard trend of the maximum Von Mises stress except for the gable angle. More specifically, the maximum stress decreased with the increase in the number of teeth, the increase in the root radius, and the increase in half pitches. On the contrary, maximum stress slightly increased with the increase in the gable angle before dropping abruptly above $a_g = 3.5°$.

Author Contributions: Conceptualization, N.T. and C.E.; methodology, N.T. and C.E.; software, N.T., C.E. and I.T.; validation, C.E. and N.T.; investigation, C.E. and I.T.; writing—original draft preparation, C.E.; writing—review and editing, C.E. and N.T.; supervision, N.T. All authors have read and agreed to the published version of the manuscript.

Funding: This research received no external funding.

Data Availability Statement: The data presented in this study are available on request from the corresponding author.

Conflicts of Interest: The authors declare no conflict of interest.

Nomenclature

D_i	Inner curvic diameter	mm
D_o	Outer curvic diameter	mm
Z	Number of teeth	-
hp	Number of half pitches	-
w	Face width	mm
R_w	Grinding wheel radius	mm
h_a	Tooth addendum	mm
h_d	Tooth dedendum	mm
a_n	Normal pressure angle	°
a_c	Chafmer angle	°
h_c	Chafmer height	mm
a_g	Gable angle	°
r_t	Tooth root radius	mm
F_s	Separating force	N
T	Torque	N·mm
A	Mean radius of the coupling	mm
F_c	Clamping force	N
σ_t	Stress due to torque	MPa
h_0	Contact height	mm
σ_c	Stress due to clamping force	MPa
σ_{eq}	Equivalent stress	MPa
$\sigma_{v,max}$	Maximum Von Mises stress	MPa
σ_y	Yield strength	MPa

References

1. Boyce, M.P. (Ed.) 18—Couplings and Alignment. In *Gas Turbine Engineering Handbook*, 4th ed.; Butterworth-Heinemann: Oxford, UK, 2012; pp. 693–719. [CrossRef]
2. Gleason Works. Curvic Coupling Design. *Gear Technol. November/December* **1986**, 34–48. Available online: https://www.geartechnology.com/ext/resources/issues/1186x/Back-to-Basics.pdf (accessed on 11 July 2023).
3. Richardson, I.J.; Hyde, T.H.; Becker, A.A.; Taylor, J.W. A comparison of two and three dimensional finite element contact analysis of Curvic couplings. *Trans. Eng. Sci.* **1999**, *24*, 11.

4. Richardson, I.; Hyde, T.; Becker, A.A.; Taylor, J. A validation of the three-dimensional finite element contact method for use with Curvic couplings. *Proc. Inst. Mech. Eng. Part G J. Aerosp. Eng.* **2002**, *216*, 63–75. [CrossRef]
5. Rencis, J.J.; Pisani, S.R. Using three-dimensional CURVIC®contact models to predict stress concentration effects in an axisymmetric model. *WIT Trans. Model. Simul.* **2005**, *39*, 10.
6. Jiang, X.-J.; Zhang, Y.-Y.; Yuan, S.-X. Analysis of the contact stresses in curvic couplings of gas turbine in a blade-off event. *Strength Mater.* **2012**, *44*, 539–550. [CrossRef]
7. Croccolo, D.; Agostinis, M.; Fini, S.; Olmi, G.; Robusto, F.; Vincenzi, N. On Hirth Ring Couplings: Design Principles Including the Effect of Friction. *Actuators* **2018**, *7*, 79. [CrossRef]
8. Zhang, D.; Yang, C.; He, T.; Liu, J.; Hong, J. Modelling and stress analysis for double-row curvic couplings. *Proc. Inst. Mech. Eng. Part C J. Mech. Eng. Sci.* **2020**, *235*, 4231–4243. [CrossRef]
9. Jung, Y.-S.; Gao, J.-C.; Lee, G.-I.; Jung, K.-R.; Kim, J.-Y. Large Curvic Coupling Gear for Ultraprecision Angle Division Using FEM. *Int. J. Precis. Eng. Manuf.* **2021**, *22*, 495–503. [CrossRef]
10. Nielson, B.J. Digital Inspection of Fixed Curvic Coupling Contact Pattern. Master's Thesis, Faculty of California Polytechnic State University, San Luis Obispo, CA, USA, 2012.
11. Efstathiou, C.; Tapoglou, N. Simulation of spiral bevel gear manufacturing by face hobbing and prediction of the cutting forces using a novel CAD-based model. *Int. J. Adv. Manuf. Technol.* **2022**, *122*, 3789–3813. [CrossRef]
12. Tapoglou, N. Calculation of non-deformed chip and gear geometry in power skiving using a CAD-based simulation. *Int. J. Adv. Manuf. Technol.* **2019**, *100*, 1779–1785. [CrossRef]
13. Li, A.-M.; Cui, H.-T.; Wen, W.-D.; Huang, F. Design and optimization of curvic coupling with double circular-arc root fillet in aero-engine. *J. Propuls. Technol.* **2016**, *37*, 146–155.
14. Huang, D.; Wang, Z.; Zeng, T. Manufacturing method for fixed curvic coupling. *China Mech. Eng.* **2013**, *24*, 1877–1880+1885.
15. Pisani, S.R.; Rencis, J.J. Investigating CURVIC coupling behavior by utilizing two- and three-dimensional boundary and finite element methods. *Eng. Anal. Bound. Elem.* **2000**, *24*, 271–275. [CrossRef]
16. Yu, Y.; Lee, B.; Cho, Y. Analysis of contact and bending stiffness for Curvic couplings considering contact angle and surface roughness. *Proc. Inst. Mech. Eng. Part E J. Process Mech. Eng.* **2019**, *233*, 1257–1267. [CrossRef]
17. Kim, B.J.; Oh, J.; Palazzolo, A. An improved preloaded Curvic coupling model for rotordynamic analyses. *J. Sound. Vib.* **2023**, *511*, 117391. [CrossRef]
18. Yang, C.; Zhang, D.; Dou, Y.; Hong, J. Stiffness Modelling for One Curvic Coupling Considering Contact Details. In Proceedings of the 14th International Conference on Vibration Problems, Crete, Greece, 1–4 September 2019; Sapountzakis, E.J., Banerjee, M., Biswas, P., Inan, E., Eds.; Springer Nature Singapore: Singapore, 2021; pp. 593–613.
19. Liu, H.; Hong, J.; Ruan, S.; Li, Z.; Cheng, G. A Model accounting for Stiffness Weakening of Curvic Couplings under Various Loading Conditions. *Math. Probl. Eng.* **2020**, *2020*, 1042375. [CrossRef]
20. *DIN3972*; Bezugsprofile von Verzahnwerkzeugen für Evolventenverzahnung nach DIN 867. Köln Beuth: Berlin, Germany, 1952.

machines

Article

Selection of Constitutive Material Model for the Finite Element Simulation of Pressure-Assisted Single-Point Incremental Forming

Ali Abdelhafeez Hassan [1,*], Gökhan Küçüktürk [2], Hurcan Volkan Yazgin [3], Hakan Gürün [4] and Duran Kaya [5]

1 Department of Engineering, School of Computing, Engineering and Digital Technologies, Teesside University, Middlesbrough TS1 3BX, UK
2 Department of Mechanical Engineering, Faculty of Engineering, Gazi University, Ankara 06560, Türkiye
3 Pi Makina, Gaziosmanpasa Mah. 97, 1 Sokak No: 6, Golbasi, Ankara 06830, Türkiye
4 Department of Manufacturing Engineering, Faculty of Technology, Gazi University, Ankara 06560, Türkiye
5 Department of Mechanical Engineering, Graduate School of Natural and Applied Sciences, Gazi University, Ankara 06560, Türkiye
* Correspondence: a.hassan@tees.ac.uk

Abstract: Pressure-assisted single-point incremental forming (PA-SPIF) is one of the emerging forming techniques for sheet metals that have been the subject of rigorous research over the past two decades. Understanding of its forming mechanisms and capabilities is growing as a result. Open gaps are still present in material constitutive modelling for accurate numerical predictions and finite-element simulations as the characteristics of localised deformation behaviour in SPIF are different from those of conventional sheet metal forming. The current investigation focused on the comparison of three different material models for the finite-element analysis of PA-SPIF of cold-rolled, dual-phase steel DP600. Experimental trials using different fluid pressures showed good agreement with simulation results with discrepancies in deformed blank thickness and shape geometry predictions of 3–11% and 10–21%, respectively. Within the tested materials and range of parameters, the fracture-forming-limit diagram (FFLD) material model was identified to be of superior accord with experiments.

Keywords: incremental forming; finite element; sheet metal; damage model; pressure assisted; hybrid manufacturing

1. Introduction

Single-point incremental forming (SPIF) is a near-net-shape manufacturing process in which a simple tool follows a specific tool path to deform sheet metal into complex prototype parts with sufficient precision. It is considered as one of the hybrid manufacturing techniques that use CNC machine tools to deform metal sheet. The process has a high potential economic payoff for the production of complex sheet metal parts in small quantities while using simple tooling. SPIF is characterised by its high formability compared with conventional sheet-metal-forming techniques. The enhanced formability is mainly attributed to the localised forming, which changes the damage and fracture behaviours of metal sheet compared with conventional processes such as stamping and deep drawing [1]. While enhanced formability was reported for high-speed forming techniques such as electromagnetic forming [2–4], the superiority of SPIF lies in the precise control of deformed shape with simple tooling. Due to the lengthy manufacturing time per part, the SPIF method is best suited for prototypes or limited production runs. In the last 20 years, specifically, the SPIF approach has been a fascinating area of study for shaping lightweight materials in the aerospace and automotive sectors [5–9]. Different domains prefer SPIF procedures, notably in the automobile industry. Important biological uses, such as cranial implants, have grown in recent years. The form of the human body must be taken into consideration when individually manufacturing medical implants for patients, which can be considered by utilising the advancement in medical imaging and CAD/CAM technologies. As a result, one of the SPIF applications that have received the greatest investigation is the creation of specialised medical implants such as skull plates, knee prostheses, face implants, and palate prostheses [10–14].

SPIF approaches have been extensively studied as the process shapes ferrous and nonferrous metals without dies. With SPIF, excessive thinning, geometry deviation, and surface-quality deterioration occur when shaping light, high-strength materials. To employ a technology that offers such flexible manufacturing in the industry, it is vital to research numerous parameters and properly understand the boundaries of the shape. Different sheet materials are shaped with single-point, two-point, and kinematic ISF methods, according to the findings. Due to their microstructure, which consists of dispersion of hard martensite particles in a soft and formable ferrite matrix, dual-phase (DP) steels are of tremendous importance to the automobile industry. This construction offers improved durability without sacrificing formability; however, because of their complex failure behaviour, their applications are limited [15].

Ham et al. [16] established the ISF shaping and effect ratio parameters. The study employed AA 3003 aluminium sheeting. The wall angle, form depth, spindle speed, and step-down size affected shaping. The depth and diameter did not affect the shape. The material thickness and tool diameter affected the wall angle more than the step-down size did. Afonso et al. [17] shaped tunnel portions (full and half-tunnel forms) with ISF. Two-millimetre-thick 1050-H111 aluminium was used. The feed rate was 1500 mm/min, and the tool diameter was 10 mm. The form correctness and greatest wall angle were tested. Up to 68° wall angle might be achieved within 5 mm. Moayedfar et al. [18] studied the effect of forming parameters during incremental sheet forming of 316 stainless steel. The findings indicated that as the spindle speed and feed rate grew, so did the sheet stretch, until the sheet could no longer stretch and the process shifted from forming to shear thinning. At greater spindle speed and feed rates, the surface quality suffered. Manco et al. [19] studied the impact of SPIF parameters on 6082-T6 aluminium alloy sheets formability. Using design of experiments and statistical analysis, they created a model to predict the final wall thickness based on the initial sheet thickness, wall angle, tool diameter, and step-down size. Mugendiran et al. [20] evaluated the formability and wall-thickness variation of the AA5052 aluminium alloy during SPIF, and it was demonstrated that the conical component is more formable than the truncated pyramid part. Bastos et al. [21] studied the effects of the tool feed rate on the SPIF of AA 1050, DP600, DP780, and DP1000. The size, tool diameter, initial wall thickness, and lubrication were constant. According to their experiments, increasing the feed rate reduced the formability and surface quality of dual-phase materials, while aluminium blanks were not affected. Azevedo et al. [22] studied the lubrication's effect on the SPIF of AA 1050 and DP780. It was concluded that SAE 30 and AL-M grease oils had a beneficial effect on aluminium 1050 surface quality, whereas Finarol B5746 and AS-40 oils are better for DP780. They reported that the greater the hardness of the material to deform, the lower the necessary lubricant viscosity required. Ham et al. [23] studied SPIF dimensional accuracy based on the material type, thickness, form type, tool diameter, and step-down size. Conical, pyramidal, and dome-shaped aluminium 5754, 6451, and 5182 were used. Laser-scanned parts were compared with CAD designs for geometric accuracy. It was shown that the accuracy of the deformed final geometry is a function of the wall angle in addition to process parameters. Zhu et al. [24] proposed an algorithm for producing flat-walled parts that cannot be produced with conventional SPIF via adapting the tool path and sheet posture. The algorithm was reported to enhance the thickness uniformity of the final deformed parts [25]. Zhan et al. [26] introduced a new numerical–analytical model to disclose the thickness fracture mechanism and estimate the fracture limit during incremental forming. It was concluded that the fracture-forming limit with increased speed tool rotation is higher compared with that of the cases without tool rotation.

Mirnia et al. [27] used the modified Mohr–Coulomb (MMC3) ductile fracture criterion to numerically study single-point incremental forming and found that fracture begins at the surface via microcracks. It was deduced that the deformation strategy (e.g., multistage or multipaths) is vital for delaying the damage evolution in the process. Chang et al. [28] developed three-sheet incremental forming (TSIF) where the surface quality and forming limit were superior to those of the conventional process for aluminium alloys AA2024 and AA7075 sheet materials.

One of the early investigations on the FEA of incremental forming was carried out by Eyckens et al. [29]. They studied the straining behaviour in the process using various material models of AA3003-O and digital image correlation. It was concluded that process parameters dictate the way the plastic deformation occurs, i.e., by through-thickness shear or rather by bending. Henrard et al. [30] demonstrated that the type of finite element, constitutive law, and identification procedure for the material parameters influenced force prediction in FEA. Essa and Hartely [31] numerically and experimentally investigated SPIF, and it was deduced that using a backing plate reduced the unintended sheet bending near the top of the outer cone location. In addition, usage of an additional kinematic support tool and modified end tool path lessened springback and decreased the pillow effect, respectively. Esmaeilpour et al. [32] confirmed that, in FEA, choosing the suitable material model was the utmost critical parameter when modelling the process. Yan et al. [33] numerically investigated multistage SPIF optimisation of aluminium AA3003-O. It was shown that the two-stage forming technique could significantly reduce the geometrical deviation, thickness variation, and forming time. Li et al. [34] investigated the tool path optimisation of induction heating-assisted SPIF using machine learning. Material formability and surface quality were improved following the optimisation of the tool path and using machine learning; however, noticeable oxidation and alpha layer of the sheet material at heating temperature ~1040 °C were reported, and they were eliminated at ~950 °C. Frikha et al. [35] explored incremental forming of grade 2 α titanium for biomedical application (hip cup prostheses). The research suggested a novel multistep process in which a deep drawing was used for the spherical cup part, whereas incremental forming was used for the outer flange part. The low material formability was enhanced, and geometric accuracy has been improved. Wang et al. [36] proposed a novel algorithm for thickness prediction in SPIF using shape geometry and surface spline (NURBS). It was presented that the proposed algorithm is robust for predicting the thickness of formed parts when compared with experimental results and FE simulated ones. The algorithm was superior to the sine law when predicting the thicknesses of variable curvature surfaces. Pepelnjak et al. [37] depicted that reducing the computational time and effort is possible through optimising mass scaling of the material and element size of the FE mesh. The optimisation was achieved via artificial neural networks applied to the results obtained from FEA and compared with the experimental results for incremental forming of the DC04 steel.

Numerical simulation is the main tool for better understanding of the process while reducing the amount of experimental trials and the associated cost. The literature on FE analysis of pressure-assisted single-point incremental forming (PA-SPIF) is limited with none devoted for the identification of the suitable and most accurate material damage model to predict material failure during the process. Hence the current investigation focused on FE analysis of the PA-SPIF of the DP600 steel by comparing the performance of different material models in predicating workpiece deformation when contrasted to experimental trials. Combined with the material tensile stress–strain empirical relation, three damage models were compared, namely the Gurson–Tvergaard–Needleman, ductile, and fracture-forming-limit damage models. Then the material model with the highest accuracy to predict deformation was reported.

2. Materials and Methods

2.1. Experimental Work

In order to investigate the influence that the PA-SPIF process has on the formation of sheet metal, an experimental setup was created. Fluid pressure is applied in the opposite direction of the surface that is being formed by the sheet material. Through the use of this apparatus, it is possible to conduct experiments both with and without the application of pressure to the fluid. The experimental arrangement allows for the fluid pressure that is applied to the sheet material to be changed in a number of different ways. Figure 1 depicts the experimental setup that has been used and the process parameters. A MICROCUT Challenger 2412 CNC machine (maximum spindle speed of 8000 rpm and spindle power of 7.5 kW) was used for the trials. Experiments involving incremental sheet forming were conducted by utilising three different pressure settings: no pressure, 0.2 bar, and 0.4 bar. The fluid used for the trials was a mixture of water and Quakercool 7101 BFF lubricant (the ratio of water to lubricant volume was 20:1).

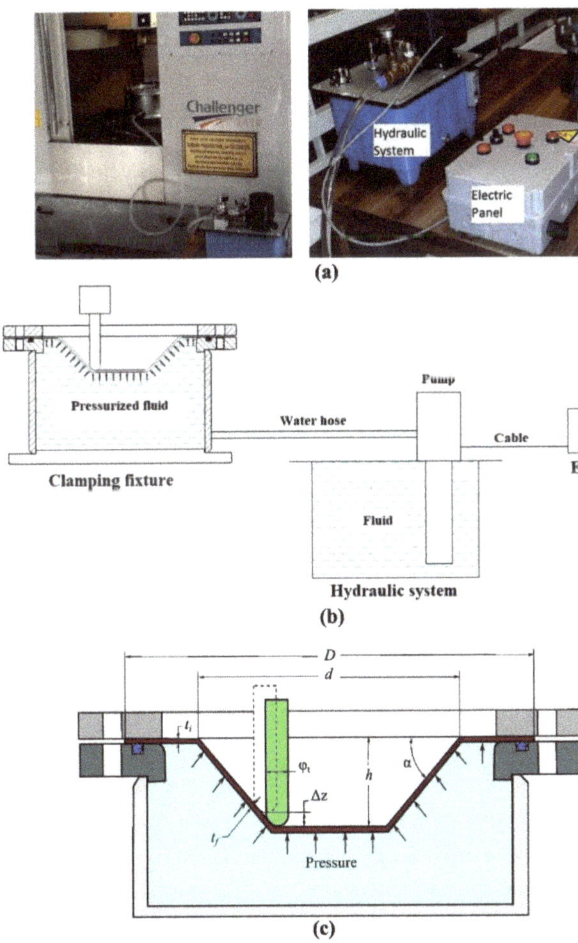

Figure 1. Experimental setup (**a**), detailed schematic of the setup (**b**), and process parameters (**c**).

Using Computer Aided Manufacturing (CAM), the processing G-codes were extracted and sent to the CNC machine in order to deform the sheet material into the desired shape. A three-axis CNC vertical milling machine was utilised to mount the experimental equipment. The tool rotation direction is clockwise, speed is 800 rpm, machining direction is counter-clockwise, step-down size is $\Delta z = 0.5$ mm, and feed rate is 1000 mm/min. The forming tool is an SAE 430B manganese bronze alloy with a 16 mm diameter and a spherical end. For sheet material with a thickness of $t_i = 0.7$ mm and a diameter of $D = 285$ mm, the forming diameter is $d = 180$ mm, form height is $h = 50$ mm, and forming angle is $\alpha = 45°$. To assure accuracy, all experiments and measurements were repeated three times, and the average of the results was calculated. Due to the high manganese content of this tool, its resistance to wear is exceptional. Figure 2 depicts the utilised forming tool, whereas Table 1 outlines its elemental composition.

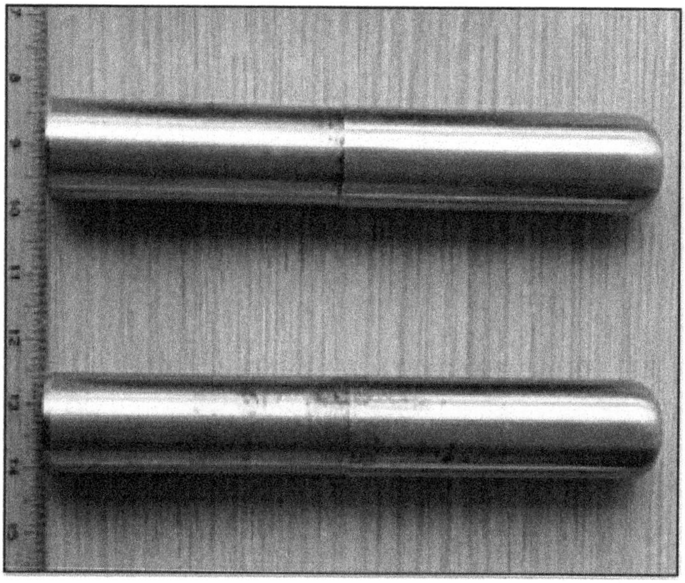

Figure 2. SAE 430B forming tools.

Table 1. Composition of forming tool and workpiece as well as properties of oil used.

Tool elemental composition (wt.%)	Al	Fe	Ni	Mn	Cu	Zn	Pb	Sn
	5.0	2.0	1.0	2.5	60	22	0.20	0.20
Oil physical Properties	**Density (gr/cm^3)**		**Viscosity (mm^2/s)at 40 °C**			**Flash point (°C)**		
	0.92		6			310		
Workpiece elemental composition (wt.%)	C	Mn	Si	Cr	Al	Ni	P	Cu
	0.116	1.545	0.289	0.634	0.042	0.041	0.029	0.019

Lubrication is required in the SPIF process to reduce friction-induced wear, enhance formability, and prolong tool life [38]. Before the experimental research, trial studies with various oil materials were conducted. According to studies on lubrication found in the scientific literature, the oils recommended for DP series materials have a low viscosity [39]. For this reason, research was pursued using a less viscous ester-based metal cutting oil. The oil of the brand CONDAT Condalu 200 was selected as the lubricant. In order to maintain consistent conditions, roughly 100 cm^3 of this oil was utilised in each experiment. Table 1 provides the physical properties of the used oil. Dual-phase DP600 sheet material, which has recently become popular in the automotive industry, has been selected as a sheet material due to its high strength and superior formability. Figure 3 depicts the results of the tensile test conducted on the sheet material (thickness of 0.7 mm) used in the studies, whereas Table 1 details its chemical composition. A laser 3D scanner with the associated software used to scan the deformed sheet geometry is depicted in Figure 4 and obtains digital cloud surfaces of the workpiece. The laser scanning approach yields precise values with a resolution of 0.01 mm. The software Geomagic Verify™ was used to process the cloud surfaces and compare with the CAD of intended geometry.

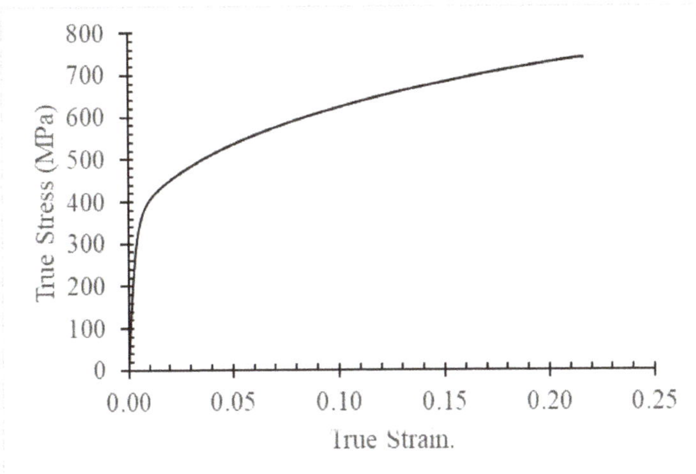

Figure 3. Experimental tensile stress–strain relation of DP600 steel (workpiece material).

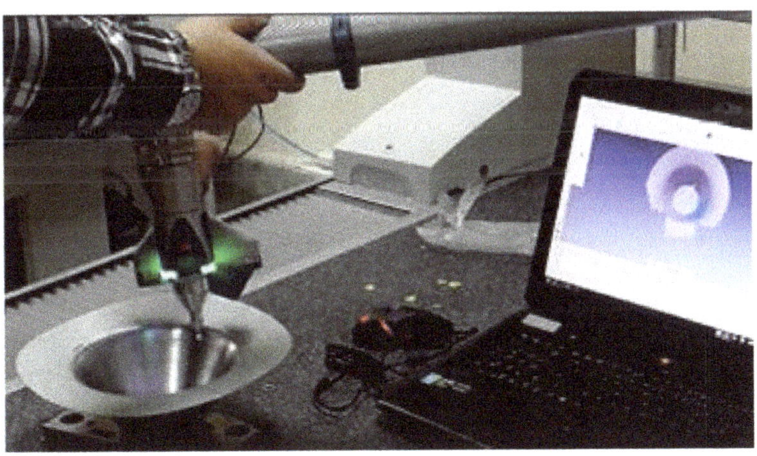

Figure 4. Laser 3D scanner with its associated software.

2.2. FE Modelling Work

Finite-element analysis of PA-SPIF of DP600 steel blanks was carried out to understand various material models' effects on deformation prediction accuracy at three fluid pressure settings. The forming tool was developed as an analytical rigid body, with only the tool hemispherical tip and 10 mm of the tool shank considered. The blank was modelled as a deformable 3D continuum homogenous shell with a thickness of 0.7 mm. Simpson's integration rule was used during the analysis with 9 integration points over the thickness to calculate the cross-sectional behaviour of the shell. The blank was meshed with S4R elements (4-node doubly curved, thin, reduced integration shell element) of 3 mm average size totalling 6820 elements. Figure 5 details the tool geometry, model assembly, and undeformed mesh of the blank.

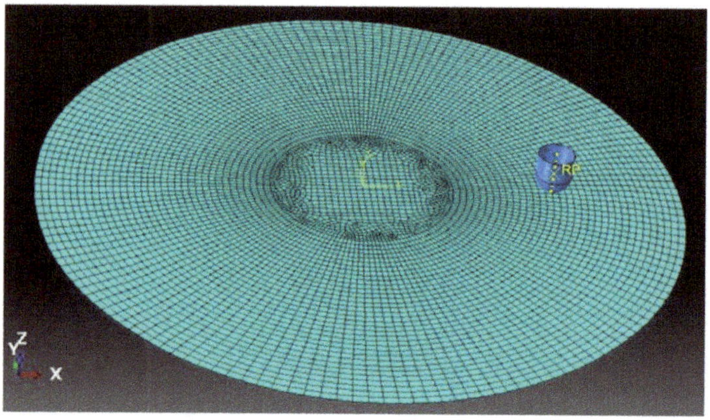

Figure 5. Model assembly and undeformed mesh.

Surface-to-surface contact was used to model the contact behaviour between the tool and blank. The friction coefficient between the tool and blank during SPIF is a function of process parameters as well as forming depth. Friction coefficient values ranging between 0.1 and 0.3 were reported during SPIF of the DP780 steel using various lubricants [22]. Therefore, an average friction coefficient value of 0.2 was considered for the tangential contact behaviour between the tool and blank in the model. The tool path used for experimental trials was inputted to the model as a displacement boundary condition of the tool to deform the blank to the required geometry. The outer perimeter of the blank was fixed, and the bottom outer surface (ring-shaped) of the blank was restricted from moving in the Z direction to represent the die and clamp plate restrictions on the blank. The fluid pressure was applied as a pressure load on the bottom surface of the blank. A dynamic explicit integration scheme was used with automatic time-increment calculation. The step total time period was 160 s, and no mass or time scaling was used. The blank material density is 7870 kg/m^3, while its Poisson's ratio and modulus of elasticity are 0.29 and 207 GPa, respectively. The material plastic behaviour obtained from experimental tensile tests was inputted to the model as tabulated values of plastic stress–strain paired values. Three material damage models were considered, namely the Gurson–Tvergaard–Needleman (GTN) damage model, ductile damage model, and fracture-forming-limit damage (FFLD) model. The GTN yield function is defined as in Equation (1).

$$\left(\frac{\sigma_{eq}}{\sigma_y}\right)^2 + 2q_1 f^*.\cosh\left(\frac{3q_2\sigma_m}{2\sigma_y}\right) - \left(1 - q_3 f^{*2}\right) = 0 \quad (1)$$

where σ_{eq} and σ_m are the equivalent von Mises stress and hydrostatic stress, respectively. q_1, q_2, and q_3 are the material parameters, and σ_y is the flow/yield stress of the material. The volume void fraction is modified to f^* due to the accelerating effects of the void coalescence as follows:

$$f^* = \begin{cases} f & , f \le f_c \\ f_c + \frac{\frac{1}{q_3}\left(q_1 + \sqrt{q_1^2 - q_3}\right) - f_c}{f_f - f_c}(f - f_c) & , f_c < f < f_f \end{cases} \quad (2)$$

where fc and f_f are the critical void volume fraction at the onset of voids coalescence and the critical void volume fraction at the onset of failure, respectively. The change in the void volume fraction is due to the enlargement of existing voids and the nucleation of new voids. Thus, the rate of change in the void volume fraction is expressed as a sum of the rate of growth of existing voids (f_G) and the rate of void nucleation (f_N) and can be defined as in Equations (3) and (4).

$$\dot{f}_G = (1 - f^*).\dot{\varepsilon}^p{}_{kk} \tag{3}$$

$$\dot{f}_N = \frac{f_N}{S_N\sqrt{2\pi}} \exp\left(-\frac{1}{2}\left(\frac{\varepsilon^p_{eff} - \varepsilon_N}{S_N}\right)\right).\dot{\varepsilon}^p_{eff} \tag{4}$$

where $\dot{\varepsilon}^p{}_{kk}$ and $\dot{\varepsilon}^p_{eff}$ are the plastic hydrostatic strain rate and effective plastic strain rate, respectively. ε_N represents the mean nucleation strain with a standard deviation of S_N, and f_N is the void volume fraction of the nucleating voids. The DP600 steel parameters for the GTN model were collected from the literature [40] as detailed in Table 2.

Table 2. GTN model parameters [40].

q_1	q_2	q_3	f_0	f_c	f_f	f_N	S_N	ε_N
1.5	1	2.25	0.008	0.15	0.25	0.00062	0.1283	0.5421

The ductile damage (DD) criterion is a phenomenological model for predicting the onset of damage due to nucleation, growth, and coalescence of voids. The ductile damage model assumes that the fracture strain is a function of stress triaxiality (η), strain rate, and temperature. Stress triaxiality is a function of the hydrostatic stress (σ_h) and the equivalent von Mises stress (σ_{eq}) as defined by equation 5. The experimentally measured values of the fracture initiation strain at different triaxialities [41] are depicted in Figure 6; the associated data were inputted to the model as tabulated values, and a strain rate of 1 s^{-1} was considered to be a suitable average value for incremental-forming processes [42]. The aforementioned fracture strain represents the start or onset of material damage (strain at the start of fracture), which is followed by material damage evolution. A generalised model that relates triaxiality to equivalent initial fracture strain ($\bar{\varepsilon}_0^{pl}$) is depicted in Equation (6), with C_1 and C_2 as the material constants, and η_0 is generally approximated to be 0.333 [43].

$$\eta = \frac{\sigma_h}{\sigma_{eq}} = \frac{\frac{1}{3}(\sigma_1 + \sigma_2 + \sigma_3)}{\sqrt{\frac{1}{2}\left((\sigma_1 - \sigma_2)^2 + (\sigma_2 - \sigma_3)^2 + (\sigma_3 - \sigma_1)^2\right)}} \tag{5}$$

$$\bar{\varepsilon}_0^{pl} = \begin{cases} \infty & , \eta \leq -\frac{1}{3} \\ \frac{C_1}{1+3\eta} & , -\frac{1}{3} \leq \eta \leq 0 \\ C_1 + (C_2 - C_1)\left(\frac{\eta}{\eta_0}\right)^2 & , 0 \leq \eta \leq \eta_0 \\ C_2\frac{\eta_0}{\eta} & , \eta_0 \leq \eta \end{cases} \tag{6}$$

In the damage evolution stage, the yield stress softens and elasticity degradation occurs until the material reaches a complete fracture when the material damage parameter (D) reaches 1 (see Figure 7). The material damage (D) in such a stage can be modelled as a function of the material fracture energy (G_f), effective plastic displacement (u^{pl}), and yield stress (σ_y) as detailed in Equation (7). The fracture energy for the DP600 steel was reported as 106 MJ/m^2. Compared with the extrapolated stress–strain relation following damage initiation, stresses during the damage evolution are softened or reduced by a factor of 1-D. To capture the residual load-carrying capability of a cracked ductile material, a postpeak softening component of the stress–strain curve is usually included in the modelling work.

$$D = 1 - \exp\left(-\int_0^{u^{pl}}\left(\frac{\sigma_y\,\dot{u}^{pl}}{G_f}\right)\right) \tag{7}$$

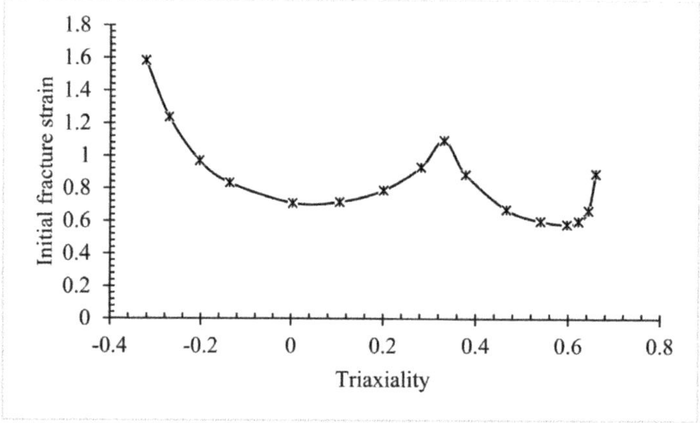

Figure 6. Fracture strain and triaxiality relation for DP600 steel.

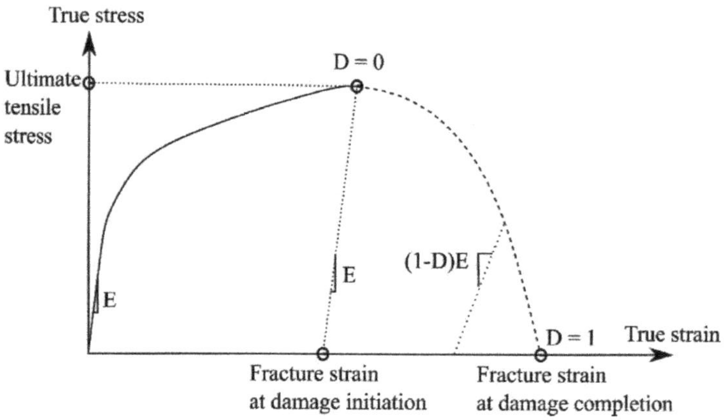

Figure 7. Typical stress–strain curve showing strain hardening and softening post damage initiation.

The basic necessity in every sheet-metal-forming operation is to evaluate the material's formability in order to achieve the desired shape without failure or fracture. Normally, the thickness of the sheet is much smaller than the other dimensions of the sheet that are responsible for selecting the plane stress condition. As a result, minor and major strains are employed to assess the formability of the material. The major and minor stresses in the necking zone are typically represented on a graph as a V-shape curve under different loading trajectories. This curve, known as the forming-limit diagram, defines the formability limit of sheet material without necking (FLD). When there is evident necking in the material during forming, FLD is applied. FLD is constructed around the strains in the necking zone. Some materials undergo instantaneous fracture without evident necking during processing; in such circumstances, formability is determined by assessing fracture strains at different loading trajectories and constructing the fracture-forming-limit diagram (FFLD). Following experimental trials, the fracture-forming-limit diagram for the DP600 sheet steel was reported by Habibi et al. [15]. The FFLD data used are depicted in Figure 8 and were entered to the model as tabulated FLD values. The damage evolution criterion (Equation (7)) was used in combination with FFLD to model the material behaviour during damage initiation and evolution.

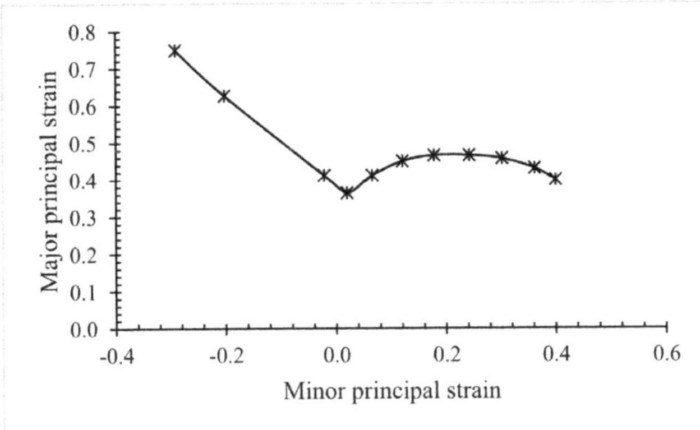

Figure 8. FFLD of DP600 steel.

3. Results and Discussion

Using laser scanning, changes in wall thickness and deviations from CAD data were measured on shaped sheet materials produced from experimental research with a resolution of 0.01 mm. Wall-thickness change and deviation readings from CAD data were measured at 10 mm intervals over the truncated cone height (from top to bottom). As demonstrated in Figure 9, PA-SPIF resulted in higher thinning than SPIF with the thinning increased when the pressure values grew. The fluid pressure acted as flexible die under the blank while it was being deformed. Therefore, the fluid pressure contributed to the squeezing or thinning of the metal sheet under the tool loading above and the fluid pressure below. This increased thinning of the blank thickness helped to reduce the springback effects by fully plasticising the sheet thickness. The springback of sheet metals in forming operations usually happens due to the residual elastic region across the sheet thickness after deformation [44]. The more plastic deformation delivered to the blank, the less springback was measured, and, thus, the less deviation from the intended final deformed geometry (CAD) of the blank. Figure 10 depicts the deviation from CAD of the final deformed blanks under different pressure settings. It is evident that, for higher pressure, the deviation is lower, which, as detailed earlier, could be attributed to higher plastic deformation across the sheet thickness with higher pressure and thus less springback or deviation. This increased plastic deformation effect was evident from the higher thinning of the blank with higher pressure.

Figure 11 depicts the deformed mesh of the model at different stages of deformation with no visible element shape distortion, which usually indicates a meshing problem by choosing the wrong element size, shape, and/or distribution. Figures 12–14 depict the von Mises stress distribution at the end of the process for different pressure settings and different material models. The material fracture (or element deletion) at the bottom or end bend of the cone, marked with a red arrow in Figure 13, was predicted by simulations that utilised a GTN material model, which was in contrast to an experimental work where no fracture was reported. However, in general, fractures usually occur near or at the bottom bend where the sidewall meets the bottom surface when forming truncated cones using SPIF. In addition, all the three material models predicted the maximum stress to be at the bottom bend. This can be attributed to the increased blank stretching and thickness thinning with tool advancement to the bottom surface to form the required shape. The predicted von Mises stress distribution and extremes for the three material models were dissimilar for different material models with discrepancies up to 100 MPa as shown in Figures 12–14. This highlights the importance of not only the plastic stress–strain relation but also the material damage model when simulating metal-forming processes and specially SPIF.

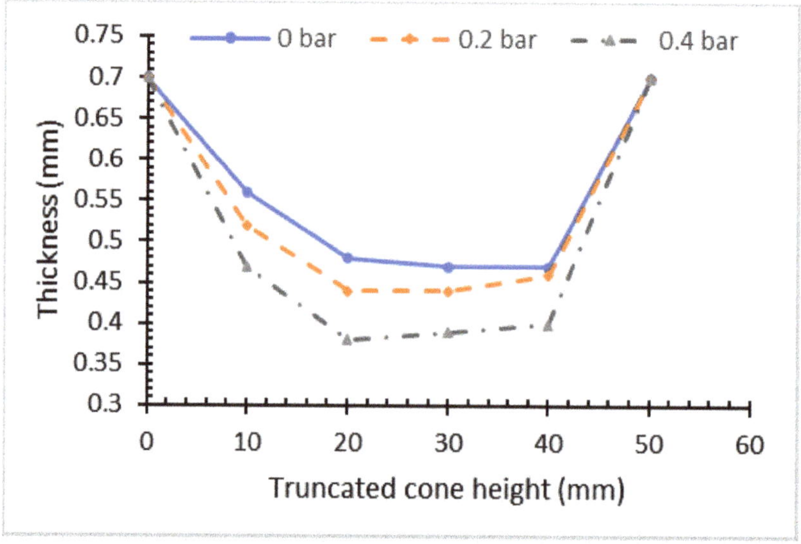

Figure 9. Thickness variation of deformed blank at different pressure settings.

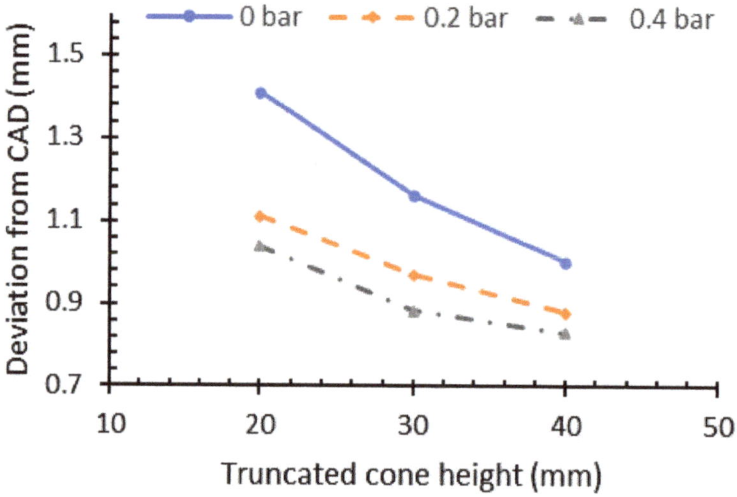

Figure 10. Deviation from CAD of deformed blanks.

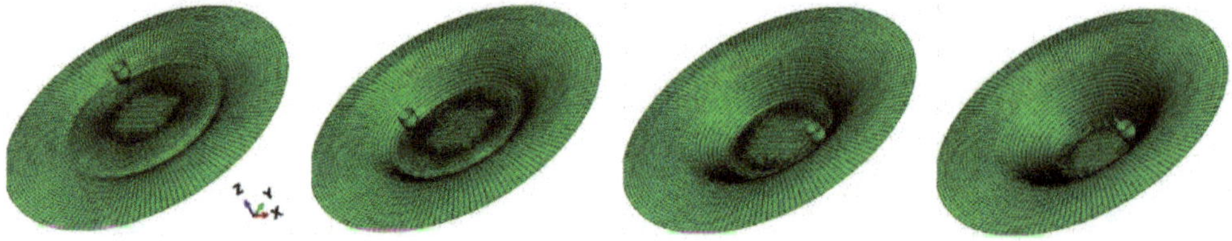

Figure 11. Blank deformation and deformed mesh over various stages (initial stage on left and final stage on right) of PA-SPIF.

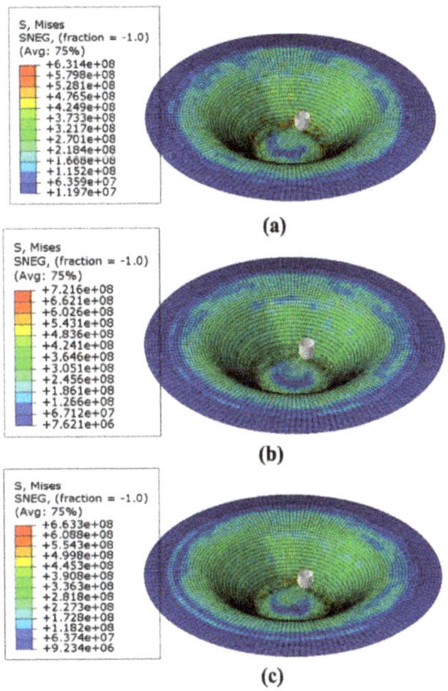

Figure 12. von Mises stress distribution for FFLD material model at (**a**) no fluid pressure, (**b**) pressure 0.2 bar, and (**c**) pressure 0.4 bar.

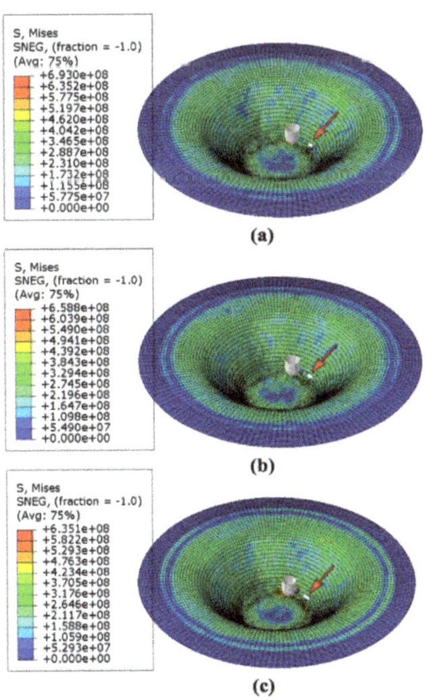

Figure 13. von Mises stress distribution for GTN material model at (**a**) no fluid pressure, (**b**) pressure 0.2 bar, and (**c**) pressure 0.4 bar. Predicted fracture regions are marked with red arrow.

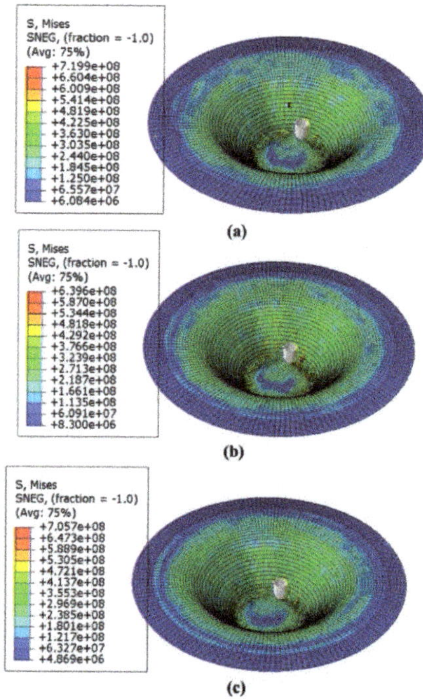

Figure 14. von Mises stress distribution for DD material model at (**a**) no fluid pressure, (**b**) pressure 0.2 bar and (**c**) pressure 0.4 bar.

Figure 15 shows the predicted final thickness of the deformed blank at different pressures and using the three material models. The data appeared to follow the same trend with experimental measured thickness where the thickness is reduced for the sidewalls of the cone compared with the top or bottom parts. The FFLD material model showed the least discrepancy from experimental work (up to 3%), whereas the GTN model showed the highest variation from measured thickness (up to 11%). This could be attributed to the drawbacks of the GTN model since it ignores fracture mechanisms brought on by shear, which might by an essential damage mechanism during SPIF [42]. Figure 16 illustrates the deviation from CAD for predicted blank deformation using the three material models, whereas Table 3 details the numerical and experimental values for each of the selected points at various cases.

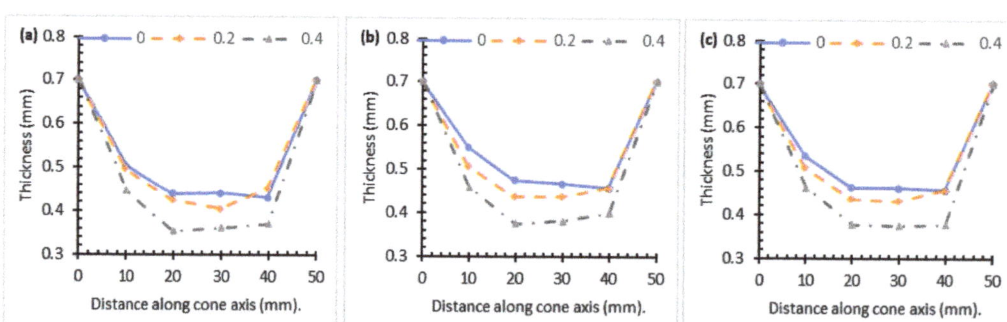

Figure 15. Thickness spatial variation at end of process as predicted by simulation using (**a**) GTN material model, (**b**) FFLD model, and (**c**) DD material model for different pressure settings.

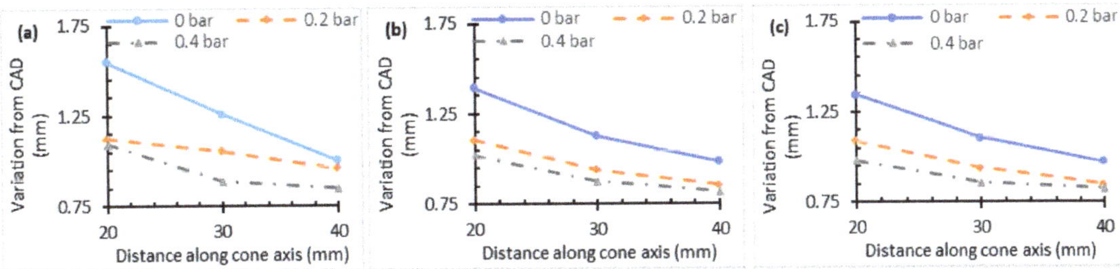

Figure 16. Variation from CAD at end of process as predicted by simulation using (**a**) GTN material model, (**b**) FFLD model, and (**c**) DD model for different pressure settings.

Table 3. Comparison of experimental and numerical results of various material models at different pressure settings.

Axial Depth (mm)	Experimental Results			FFLD			GTN			DD		
	0.0 bar	0.2 bar	0.4 bar	0.0 bar	0.2 bar	0.4 bar	0.0 bar	0.2 bar	0.4 bar	0.0 bar	0.2 bar	0.4 bar
	Variation from CAD (mm)											
20	1.4	1.11	1.03	1.39	1.12	1.01	1.42	1.21	1.05	1.38	1.06	0.99
30	1.1	0.96	0.88	1.16	0.95	0.86	1.17	1.04	0.89	1.13	0.93	0.86
40	1.01	0.88	0.83	0.99	0.86	0.81	1.10	0.92	0.84	0.94	0.86	0.83
	Thickness (mm)											
10	0.56	0.52	0.47	0.56	0.51	0.46	0.51	0.50	0.44	0.53	0.50	0.44
20	0.48	0.44	0.38	0.47	0.43	0.37	0.44	0.41	0.36	0.46	0.41	0.36
30	0.47	0.44	0.39	0.46	0.43	0.38	0.45	0.40	0.37	0.47	0.42	0.39
40	0.47	0.46	0.41	0.47	0.45	0.40	0.46	0.44	0.37	0.45	0.43	0.38

The FFLD material model showed the least variation with CAD, and it was within 90% agreement with experimental work, whereas the GTN and DD models agreed by 79% and 84%, respectively. The aforementioned results, which are summarised in Table 4, confirmed that the FFLD material model was the most accurate one for predicting material deformation and damage in the PA-SPIF process for DP600 steel metal sheets within the tested range of values.

Table 4. Summary of findings.

Characteristics	Material Model FFLD	DD	GTN
No-fracture prediction	Yes	Yes	No
Thickness discrepancy up to (%)	3	6	10
Variation from CAD up to (%)	10	16	21

4. Conclusions

The current investigation compared three different material models for the finite-element analysis of pressure-assisted single-point incremental forming of the cold-rolled, dual-phase steel DP600. Experimental trials using 0.2 bar and 0.4 bar fluid pressures besides no fluid pressure (0 bar) showed good agreement with the simulation results with errors in deformed blank thickness and deformed geometry predictions of 3–11% and 10–21%, respectively. Based on the tested range of parameters and materials, the FFLD material model showed the least discrepancy with experiments, whereas the GTN model depicted the highest discrepancy in predicting thickness, geometry, and material nonfracture condition. The GTN model ignores damage mechanisms due to shear, which might be vital in SPIF. The current investigation revealed the importance of a damage model in predicting deformation during SPIF using finite-element analysis; however, the plastic stress–strain constitutive model is of equal importance. Experimental uniaxial tension data were used in the current investigation for the constitutive stress–strain model; however, further testing at different temperatures, strain rates, and strain configurations (e.g., biaxial stretching) is needed for further improvements in modelling work.

Author Contributions: Conceptualisation, A.A.H.; experimental methodology, G.K. and H.G.; software, A.A.H.; validation, H.V.Y. and D.K.; investigation, A.A.H.; writing—original draft preparation and writing—review and editing, A.A.H.; visualisation, A.A.H.; project administration, G.K. All authors have read and agreed to the published version of the manuscript.

Funding: This research received no external funding.

Acknowledgments: The authors acknowledge the support given by Birmingham Environment for Academic Research (BEAR) at the University of Birmingham via access to the HPC facility, BlueBEAR, at some early stages of this research.

Conflicts of Interest: The authors declare no conflict of interest.

References

1. Martins, P.A.F.; Bay, N.; Skjoedt, M.; Silva, M.B. Theory of Single Point Incremental Forming. *CIRP Ann.-Manuf. Technol.* **2008**, *57*, 247–252. [CrossRef]
2. Abdelhafeez, A.M.; Nemat-Alla, M.M.; El-Sebaie, M.G. FEA of Electromagnetic Forming Using a New Coupling Algorithm. *Int. J. Appl. Electromagn. Mech.* **2013**, *42*, 157–169. [CrossRef]
3. Abdelhafeez, A.M.; Nemat-Alla, M.M.; El-Sebaie, M.G. Finite Element Analysis of Electromagnetic Bulging of Sheet Metals. *Int. J. Sci. Eng. Res.* **2012**, *3*, 180–187.
4. Abdelhafeez, A.M.; Nemat-Alla, M.M.; El-Sebaie, M.G. FEA of Electromagnetic Forming Using a New Coupling Algorithm: Effects of Strain Hardening Properties and Anisotropy. *Int. J. Sci. Eng. Res.* **2014**, *5*, 1069–1075.
5. Emmens, W.C.; Sebastiani, G.; van den Boogaard, A.H. The Technology of Incremental Sheet Forming—A Brief Review of the History. *J. Mater. Processing Technol.* **2010**, *210*, 981–997. [CrossRef]
6. Peter, I.; Fracchia, E.; Canale, I.; Maiorano, R. Incremental Sheet Forming for Prototyping Automotive Modules. *Procedia Manuf.* **2019**, *32*, 50–58. [CrossRef]
7. Duflou, J.R.; Habraken, A.-M.; Cao, J.; Malhotra, R.; Bambach, M.; Adams, D.; Vanhove, H.; Mohammadi, A.; Jeswiet, J. Single Point Incremental Forming: State-of-the-Art and Prospects. *Int. J. Mater.* **2018**, *11*, 743–773. [CrossRef]
8. Scheffler, S.; Pierer, A.; Scholz, P.; Melzer, S.; Weise, D.; Rambousek, Z. Incremental Sheet Metal Forming on the Example of Car Exterior Skin Parts. *Procedia Manuf.* **2019**, *29*, 105–111. [CrossRef]
9. Behera, A.K.; de Sousa, R.A.; Ingarao, G.; Oleksik, V. Single Point Incremental Forming: An Assessment of the Progress and Technology Trends from 2005 to 2015. *J. Manuf. Processes* **2017**, *27*, 37–62. [CrossRef]
10. Kumar, A.; Gulati, V.; Kumar, P.; Singh, V.; Kumar, B.; Singh, H. Parametric Effects on Formability of AA2024-O Aluminum Alloy Sheets in Single Point Incremental Forming. *J. Mater. Res. Technol.* **2019**, *8*, 1461–1469. [CrossRef]
11. McAnulty, T.; Jeswiet, J.; Doolan, M. Formability in Single Point Incremental Forming: A Comparative Analysis of the State of the Art. *CIRP J. Manuf. Sci. Technol.* **2017**, *16*, 43–54. [CrossRef]
12. Kim, T.J.; Yang, D.Y. Improvement of Formability for the Incremental Sheet Metal Forming Process. *Int. J. Mech. Sci.* **2000**, *42*, 1271–1286. [CrossRef]
13. Ambrogio, G.; De Napoli, L.; Filice, L.; Gagliardi, F.; Muzzupappa, M. Application of Incremental Forming Process for High Customised Medical Product Manufacturing. *J. Mater. Processing Technol.* **2005**, *162–163*, 156–162. [CrossRef]
14. Cheng, Z.; Li, Y.; Xu, C.; Liu, Y.; Ghafoor, S.; Li, F. Incremental Sheet Forming towards Biomedical Implants: A Review. *J. Mater. Res. Technol.* **2020**, *9*, 7225–7251. [CrossRef]
15. Habibi, N.; Ramazani, A.; Sundararaghavan, V.; Prahl, U. Failure Predictions of DP600 Steel Sheets Using Various Uncoupled Fracture Criteria. *Eng. Fract. Mech.* **2018**, *190*, 367–381. [CrossRef]
16. Ham, M.; Jeswiet, J. Single Point Incremental Forming and the Forming Criteria for AA3003. *CIRP Ann.* **2006**, *55*, 241–244. [CrossRef]
17. Afonso, D.; de Sousa, R.A.; Torcato, R. Incremental Forming of Tunnel Type Parts. *Procedia Eng.* **2017**, *183*, 137–142. [CrossRef]
18. Moayedfar, M.; Leman, Z.; bin Baharudin, B.T.H.T. Incremental Sheet Forming (ISF) of AISI 316 Stainless Steel Sheet Using CNC Milling Machine. *AMR* **2014**, *939*, 322–327. [CrossRef]
19. Manco, G.L.; Ambrogio, G. Influence of Thickness on Formability in 6082-T6. *Int. J. Mater.* **2010**, *3*, 983–986. [CrossRef]
20. Mugendiran, V.; Gnanavelbabu, A. Comparison of FLD and Thickness Distribution on AA5052 Aluminium Alloy Formed Parts by Incremental Forming Process. *Procedia Eng.* **2014**, *97*, 1983–1990. [CrossRef]

21. Pereira Bastos, R.N.; Alves de Sousa, R.J.; Fernandes Ferreira, J.A. Enhancing Time Efficiency on Single Point Incremental Forming Processes. *Int. J. Mater.* **2016**, *9*, 653–662. [CrossRef]
22. Azevedo, N.G.; Farias, J.S.; Bastos, R.P.; Teixeira, P.; Davim, J.P.; Alves de Sousa, R.J. Lubrication Aspects during Single Point Incremental Forming for Steel and Aluminum Materials. *Int. J. Precis. Eng. Manuf.* **2015**, *16*, 589–595. [CrossRef]
23. Ham, M.; Jeswiet, J. Dimensional Accuracy of Single Point Incremental Forming. *Int. J. Mater.* **2008**, *1*, 1171–1174. [CrossRef]
24. Zhu, H.; Liu, L. Research the CNC Incremental Forming of Straight-Wall Parts Based on a Virtual Auxiliary Body. *J. Mater. Processing Technol.* **2021**, *288*, 116841. [CrossRef]
25. Zhu, H.; Wang, Y.; Kang, J. Research on Combinatorial Optimization of Multidirectional Sheet Postures for Forming Thickness Uniformity. *J. Mech. Sci. Technol.* **2020**, *34*, 4251–4261. [CrossRef]
26. Zhan, X.; Wang, Z.; Li, M.; Hu, Q.; Chen, J. Investigations on Failure-to-Fracture Mechanism and Prediction of Forming Limit for Aluminum Alloy Incremental Forming Process. *J. Mater. Processing Technol.* **2020**, *282*, 116687. [CrossRef]
27. Mirnia, M.J.; Vahdani, M.; Shamsari, M. Ductile Damage and Deformation Mechanics in Multistage Single Point Incremental Forming. *Int. J. Mech. Sci.* **2018**, *136*, 396–412. [CrossRef]
28. Chang, Z.; Chen, J. Investigations on the Deformation Mechanism of a Novel Three-Sheet Incremental Forming. *J. Mater. Processing Technol.* **2020**, *281*, 116619. [CrossRef]
29. Eyckens, P.; Belkassem, B.; Henrard, C.; Gu, J.; Sol, H.; Habraken, A.M.; Duflou, J.R.; Van Bael, A.; Van Houtte, P. Strain Evolution in the Single Point Incremental Forming Process: Digital Image Correlation Measurement and Finite Element Prediction. *Int. J. Mater.* **2011**, *4*, 55–71. [CrossRef]
30. Henrard, C.; Bouffioux, C.; Eyckens, P.; Sol, H.; Duflou, J.R.; Van Houtte, P.; Van Bael, A.; Duchêne, L.; Habraken, A.M. Forming Forces in Single Point Incremental Forming: Prediction by Finite Element Simulations, Validation and Sensitivity. *Comput. Mech.* **2011**, *47*, 573–590. [CrossRef]
31. Essa, K.; Hartley, P. An Assessment of Various Process Strategies for Improving Precision in Single Point Incremental Forming. *Int. J. Mater.* **2011**, *4*, 401–412. [CrossRef]
32. Esmaeilpour, R.; Kim, H.; Park, T.; Pourboghrat, F.; Mohammed, B. Comparison of 3D Yield Functions for Finite Element Simulation of Single Point Incremental Forming (SPIF) of Aluminum 7075. *Int. J. Mech. Sci.* **2017**, *133*, 544–554. [CrossRef]
33. Yan, Z.; Hassanin, H.; El-Sayed, M.A.; Eldessouky, H.M.; Djuansjah, J.; Alsaleh, A.N.; Essa, K.; Ahmadein, M. Multistage Tool Path Optimisation of Single-Point Incremental Forming Process. *Materials* **2021**, *14*, 6794. [CrossRef] [PubMed]
34. Li, W.; Shu, C.; Hassan, A.; Attallah, M.M.; Essa, K. Application of Machine Learning on Tool Path Optimisation and Cooling Lubricant in Induction Heating-Assisted Single Point Incremental Sheet Forming of Ti-6Al-4V Sheets. *Int. J. Adv. Manuf. Technol.* **2022**. [CrossRef]
35. Frikha, S.; Giraud-Moreau, L.; Bouguecha, A.; Haddar, M. Simulation-Based Process Design for Asymmetric Single-Point Incremental Forming of Individual Titanium Alloy Hip Cup Prosthesis. *Materials* **2022**, *15*, 3442. [CrossRef] [PubMed]
36. Wang, Y.; Wang, L.; Zhang, H.; Gu, Y.; Ye, Y. A Novel Algorithm for Thickness Prediction in Incremental Sheet Metal Forming. *Materials* **2022**, *15*, 1201. [CrossRef] [PubMed]
37. Pepelnjak, T.; Sevšek, L.; Lužanin, O.; Milutinović, M. Finite Element Simplifications and Simulation Reliability in Single Point Incremental Forming. *Materials* **2022**, *15*, 3707. [CrossRef]
38. Mulay, A.; Ben, S.; Ismail, S. Lubricant Selection and Post Forming Material Characterization in Incremental Sheet Forming. *IOP Conf. Ser. Mater. Sci. Eng.* **2020**, *967*, 012072. [CrossRef]
39. Diabb, J.; Rodríguez, C.A.; Mamidi, N.; Sandoval, J.A.; Taha-Tijerina, J.; Martínez-Romero, O.; Elías-Zúñiga, A. Study of Lubrication and Wear in Single Point Incremental Sheet Forming (SPIF) Process Using Vegetable Oil Nanolubricants. *Wear* **2017**, *376–377*, 777–785. [CrossRef]
40. Isik, K.; Gerstein, G.; Gutknecht, F.; Clausmeyer, T.; Nürnberger, F.; Maier, H.J.; Tekkaya, A.E. Investigations of Ductile Damage in DP600 and DC04 Deep Drawing Steel Sheets during Punching. *Procedia Struct. Integr.* **2016**, *2*, 673–680. [CrossRef]
41. Heibel, S.; Dettinger, T.; Nester, W.; Clausmeyer, T.; Tekkaya, A. Damage Mechanisms and Mechanical Properties of High-Strength Multiphase Steels. *Materials* **2018**, *11*, 761. [CrossRef] [PubMed]
42. Gatea, S.; Ou, H.; Lu, B.; McCartney, G. Modelling of Ductile Fracture in Single Point Incremental Forming Using a Modified GTN Model. *Eng. Fract. Mech.* **2017**, *186*, 59–79. [CrossRef]
43. Yu, H.L.; Jeong, D.Y. Application of a Stress Triaxiality Dependent Fracture Criterion in the Finite Element Analysis of Unnotched Charpy Specimens. *Theor. Appl. Fract. Mech.* **2010**, *54*, 54–62. [CrossRef]
44. Marciniak, Z.; Duncan, J.L.; Hu, S.J. *Mechanics of Sheet Metal Forming*, 2nd ed.; Butterworth-Heinemann: Oxford, UK, 2002; ISBN 978-0-7506-5300-8.

MDPI AG
Grosspeteranlage 5
4052 Basel
Switzerland
Tel.: +41 61 683 77 34

Machines Editorial Office
E-mail: machines@mdpi.com
www.mdpi.com/journal/machines

www.ingramcontent.com/pod-product-compliance
Lightning Source LLC
Chambersburg PA
CBHW061551210926
51850CB00009B/2068